Surface Engineering - Foundational Concepts, Techniques and Applications

Edited by Jian Wang

Published in London, United Kingdom

Surface Engineering - Foundational Concepts, Techniques and Applications
http://dx.doi.org/10.5772/intechopen.1004978
Edited by Jian Wang

Contributors
Jialiang Qi, Jian Wang, Jiaxiang Ren, Lei Zhao, Mahlatse S. Rabothata, Nthape P. Mphasha, Peng Cheng, Simon C. Tung, Timothy Ryan Dunne, Wen Shuai

First published in London, United Kingdom, 2025 by IntechOpen
IntechOpen is the global imprint of INTECHOPEN LIMITED, registered in England and Wales, registration number: 11086078, 167-169 Great Portland Street, London, W1W 5PF, United Kingdom

For EU product safety concerns: IN TECH d.o.o., Prolaz Marije Krucifikse Kozulić 3, 51000 Rijeka, Croatia, info@intechopen.com or visit our website at intechopen.com.

British Library Cataloguing-in-Publication Data
A catalogue record for this book is available from the British Library

Surface Engineering - Foundational Concepts, Techniques and Applications
Edited by Jian Wang
p. cm.

This title is part of the Materials Science Book Series, Volume 7
Topic: Surface Science
Series Editor: Chonghe Li
Topic Editor: Wan Sharuzi Wan Harun

Print ISBN 978-1-83634-143-7
Online ISBN 978-1-83634-142-0
eBook (PDF) ISBN 978-1-83634-144-4
ISSN 3049-8856

If disposing of this product, please recycle the paper responsibly.

IntechOpen Book Series

Materials Science

Volume 7

Aims and Scope of the Series

Materials science has always occupied an extremely high position in the human development process. As we explore the oceans of stars, various industries have put forward more stringent requirements for the performance of materials, forcing us to pay more and more attention to the development of new materials. At the same time, the formation of a data-driven scientific paradigm is dramatically shortening the development cycle of new materials. The huge data generated by synergistically combining theories, high-throughput experiments, high-throughput computation, and artificial intelligence is greatly contributing to our ability to utilize materials science to solve real-world problems. The three topics of this book series - Metals and Nonmetals; Composite Materials; and Surface Science - will address important areas of advancement in materials science. There will be a range of interesting works published under these topics.

Meet the Series Editor

Prof. Chonghe Li received his Ph.D. from the Chinese Academy of Sciences in 1995. From 1995 to 2000, he worked as a researcher at the Shanghai Institute of Metallurgy, Chinese Academy of Sciences, where he also served as director of the research laboratory. In 2000, he was appointed Professor at the Institute of High-Performance Computing in Singapore, where he worked on computation and simulation in materials science until 2004. Since then, he has been a professor at the School of Materials Science and Engineering, Shanghai University, China, as well as the director of the Shanghai Specialty Casting Engineering and Technology Research Center. Prof. Li's research focuses on titanium alloys, titanium-aluminum single crystals, intermetallic compounds, theoretical calculations, alloy design, and special refractory materials. His broad scientific expertise is well recognized by the scientific community around the world. He is a member of the editorial board of the journal Metals. As an author, he has published more than 200 peer-reviewed papers, 2 books, and over 40 patents.

Meet the Volume Editor

Jian Wang, a professor at Beijing University of Chemical Technology since 2018, specializes in polymer engineering. He earned a Ph.D. in Mechanical Engineering (2010) and has researched at the University of Wisconsin–Madison (2009-2010), Beijing Institute of Technology (2010-2018), and RWTH Aachen University (2017-2020). Prof. Wang has been honored as an experienced researcher by the Alexander von Humboldt Foundation, and he is a fellow of the China Plastic Processing Industry Association (CPPIA) and the China Plastics Machinery Industry Association (CPMIA). He has published over 100 peer-reviewed papers and 70 patents. He has led numerous research projects funded by the Natural Science Foundation of China, the National Key Technology Research and Development Program of China, and various university and enterprise initiatives.

Contents

Preface

The rapid advancement of industrial modernization has placed increasingly stringent demands on the surface performance of mechanical components—particularly those operating under extreme conditions, such as high speeds, elevated temperatures, heavy loads, and corrosive environments. Material failure often originates at the surface in the form of wear, corrosion, or oxidation, eventually leading to component degradation and costly equipment downtime. Enhancing surface properties through targeted engineering solutions not only extends service life but also improves productivity, conserves resources, and reduces environmental impact. Surface engineering is a multidisciplinary field that focuses on modifying the surface properties of metals and non-metals through coatings, treatments, or hybrid techniques, altering morphology, composition, microstructure, and stress states to achieve performance that is unattainable by the base material alone. This approach enables the creation of tailored surface characteristics without compromising the integrity of the bulk material.

The edited book *Surface Engineering – Foundational Concepts, Techniques and Applications* provides a comprehensive exploration of advanced surface modification methods, their industrial applications, and the challenges posed by corrosion and wear. Chapter 1, "Introductory Chapter: Unveiling Present and Future Horizons of Surface Engineering", traces the evolution of the field, outlining its foundational principles, technological classifications, and critical industrial roles. Chapter 2, "Advanced Surface Modification Techniques", refers to a diverse range of surface treatment processes and techniques, with a focus on laser surface treatment (LST), plasma surface treatment (PST), ion implantation, and electron beam surface treatment (EBST). An overview of the effects of surface modification on mechanical and tribological properties is outlined. The chapter also describes the prospects and challenges of ASM in the aerospace, automotive, and medical fields. Moving on to several specific aspects of surface engineering applications, Chapter 3, "Powertrain Tribology Development Trends and Impact of Surface Engineering on Friction and Wear Control", presents the fundamentals of surface coatings or textures specific to the environments of powertrain components, together with discussions on the impact of surface engineering on friction and wear. Deploying carbon capture, utilization, and storage (CCUS), Chapter 4, Surface Engineering Solutions for Corrosion Protection in CCUS Tubular Applications", provides a comprehensive review of these technologies in downhole tubular applications, covering their history, development routes, failure analysis, current issues, field experiences, and selection criteria. The focus on industrial applications becomes even more prominent in Chapter 5, "Application of Surface Engineering in Metal and Plastic Joining of Insert Injection Molding". This chapter explores the application of surface engineering techniques to address the challenges of metal and plastic joining in insert injection molding. The chapter explores various surface engineering techniques and their mechanisms, presenting their practical applications to enhance the adhesion between metals and plastics.

By synthesizing insights from diverse industries, this book equips researchers, engineers, and professionals in materials science, chemistry, physics, and engineering with both theoretical frameworks and practical solutions.

I would like to extend my heartfelt thanks to all contributors whose insights and expertise have greatly enriched this book. Special thanks go to the editorial team, whose efforts have been instrumental in bringing this collection to fruition. I would also like to express my deepest gratitude to my wife and son, whose unwavering support and understanding during the editing process have been invaluable. Their patience and encouragement have allowed me to dedicate the time and effort needed to bring this book to completion.

Jian Wang
Beijing University of Chemical Technology,
Beijing, China

Chapter 1

Introductory Chapter: Unveiling Present and Future Horizons of Surface Engineering

Jian Wang

1. Introduction

Surface engineering is a scientific and technological discipline focused on enhancing functional characteristics such as wear resistance, corrosion resistance, oxidation resistance, and biocompatibility by modifying the composition, structure, or properties of material surfaces or near-surface regions. The origins of surface engineering trace back to ancient practices, including historical gold leaf application, gilding techniques, quenching technology, and tung oil paint anti-corrosion methods—all early manifestations of surface performance optimization. The formal conceptualization of surface engineering emerged in the modern era. In 1982, Bell advocated for officially adopting the term "surface engineering" at the International Federation for Heat Treatment and Surface Engineering (IFHTSE) conference. By 1986, the discipline was defined as "a systematic methodology for improving material functionality through surface optimization," establishing it as an independent field [1] encompassing coating deposition, surface modification, performance testing, and other related technologies.

Surface engineering stands at the forefront of materials science, bridging the gap between theoretical research and industrial innovation. As industries push the boundaries of performance, durability, and sustainability, the role of surface engineering has become indispensable. By modifying the outermost layers of materials without altering their bulk properties, this field enhances resistance to wear, corrosion, fatigue, and even biofouling, enabling breakthroughs across aerospace, automotive, biomedical, and energy sectors.

Surface engineering technologies can be categorized into physical and chemical approaches. Physical methods employ mechanical forces, thermal energy, or photonic energy for surface morphology control, including laser processing [2] (micron-scale etching), shot peening (wear resistance enhancement), and embossing techniques (microstructure formation). These techniques find applications in optical components, microfluidic chips, and high-precision mold manufacturing. Chemical approaches rely on chemical reactions to modify surface composition and functionality. Examples include chemical vapor deposition [3] (CVD), photochemical etching [4, 5] (enabling selective patterning), and self-assembled monolayers (SAMs). These methods are widely applied in corrosion-resistant coatings, biosensors, and superhydrophobic surfaces.

2. Purpose and goals

In synthesizing cross-disciplinary knowledge, this book establishes an innovative nexus between fundamental research and technological applications in surface engineering. Grounded in the convergence of material science, chemistry, physics, and engineering principles, we systematically construct a knowledge framework that is aligned with emerging technological demands. The purpose is to explore the critical role of surface engineering in enhancing material performance, durability, and functionality across various industrial applications. Surface modification techniques and advanced coatings have become indispensable in addressing challenges related to wear, corrosion, friction, and adhesion in demanding environments. By examining cutting-edge research and practical implementations, this section aims to provide a comprehensive understanding of how surface engineering innovations drive progress in key technological areas. Advanced surface modification techniques present the latest advancements in surface treatment technologies, including laser texturing, plasma coatings, and chemical functionalization, highlighting their impact on material properties and industrial applicability. The application of surface engineering in metal and plastic joining for insert injection molding discusses how surface engineering improves adhesion, bonding strength, and longevity in hybrid structures, particularly in automotive and consumer product manufacturing. Emerging trends in tribological coatings and surface treatments that enhance efficiency, reduce friction, and extend the lifespan of powertrain components are analyzed. Protective coatings and surface treatments designed to mitigate corrosion in carbon capture, utilization, and storage (CCUS) systems are evaluated, ensuring reliability in harsh operational conditions. This book serves as a vital resource for academics, engineers, and industry professionals seeking to deepen their understanding of surface engineering and leverage its innovation potential.

We invite researchers, engineers, and students to explore the chapters ahead, each offering a unique perspective on how surface engineering continues to push the limits of material performance. By fostering collaboration between science and industry, we can unlock new frontiers in durability, efficiency, and environmental responsibility, ensuring that surface engineering remains a cornerstone of technological progress for decades to come.

3. Conclusion

Surface engineering is a multidisciplinary field focused on modifying the surface properties of materials to enhance performance, durability, and functionality. It involves altering the surface layer of a material while maintaining its bulk properties, improving resistance to wear, corrosion, fatigue, and other surface-related failures. Surface engineering plays a critical role in industries where material failure starts at the surface, making it essential for innovation in manufacturing, energy, healthcare, and more. This book aims to illuminate these advancements, providing readers with insights into both present achievements and future horizons.

As we stand on the cusp of breakthroughs in nanotechnology, smart coatings, and sustainable surface treatments, the possibilities for innovation are boundless. Future advancements emphasize physicochemical synergies, such as laser-induced deposition for intelligent responsive surfaces or bio-inspired structures integrated with nanocoatings to enhance multifunctionality. These innovations are driving breakthroughs in energy, healthcare, and micro-nano manufacturing sectors. Surface engineering not only addresses bottlenecks in high-end materials for fields such as aerospace, biomedical applications, and energy equipment, but also drives sustainable industrial development through its "low-cost, low-energy consumption" strategy.

Author details

Jian Wang
Beijing University of Chemical Technology, China

*Address all correspondence to: wjj_0107@163.com

References

[1] Bell T. Surface engineering: A key to improved performance of materials. Surface Engineering. 1986;**2**(1):1-4

[2] Dumitru ED, Romano S. Laser microstructuring of steel surfaces for tribological applications. Applied Physics A: Materials Science & Processing. 2000;**70**(4):485-488

[3] Vaeth KM, Jackman RJ, Black AJ, Whitesides GM, Jensen KF. Use of microcontact printing for generating selectively grown films of poly(p-phenylene vinylene) and parylenes prepared by chemical vapor deposition. Langmuir. 2000;**16**(21):8495-8500

[4] Tehrani AF, Imanian E. A new etchant for the chemical machining of St304. Journal of Materials Processing Technology. 2004;**149**(1-3):404-408

[5] Bruzzone AAG, Costa HL, Lonardo PM, Lucca DA. Advances in engineered surfaces for functional performance. CIRP Annals - Manufacturing Technology. 2008;**57**(1):750-769

Chapter 2

Advanced Surface Modification Techniques

Nthape P. Mphasha and Mahlatse S. Rabothata

Abstract

Advanced surface modification (ASM) refers to the diverse range of surface treatment processes and techniques used to change the surface properties of materials, thereby enhancing their performance, durability, and functionality for various industrial applications. The surface of materials plays a critical role in determining their overall behaviour and application suitability. This chapter reviews the principles of various ASM techniques, with a focus on laser surface treatment (LST), plasma surface treatment (PST), ion implantation, and electron beam surface treatment (EBST). An overview of the effects of surface modification on mechanical and tribological properties is outlined. The chapter also describes the prospects and challenges of ASM in the aerospace, automotive, and medical fields. Despite the appreciable adoption and application of ASM techniques in various industries, several challenges persist that limit their full potential. It is envisioned that the integration of multiple surface treatments or hybrid surface treatments could provide an opportunity in advancing the development and effective application of ASM techniques. However, the successful application of these techniques is reliant on the understanding of processing parameters in relation to diverse materials as well as challenges innate to each technique.

Keywords: surface modification, laser surface treatment, plasma surface treatment, ion implantation, electron beam treatment

1. Introduction

Contemporary industries require materials with exceptional properties that can meet the stringent requirements of competitive markets. This need drives continuous innovation and development in materials science and engineering. Metals and their alloys, ceramics, and polymers, fundamental to numerous industries, are particularly critical, with their utility ever evolving alongside the changing technological, economic, and environmental demands. Surface modification and advanced surface modification (ASM) both involve the use of surface engineering techniques to alter and improve the surface properties of materials without necessarily changing their bulk properties. However, the extent, techniques, and complexity differentiate them. For instance, basic or simple surface modification techniques involve relatively straightforward techniques (i.e., polishing and coating) to change the surface properties. Advanced surface modification techniques, on the other hand, involve more sophisticated and often multistep processes that provide enhanced or novel functionalities to the surface. These techniques are usually more accurate and can be tailored to specific applications requiring high performance. Advanced surface modification is critical in applications where surface interactions, such as friction, adhesion, wear, and corrosion, are the primary factors limiting the material's performance. This is important in the automotive, aerospace, and biomedical industries. For example, the ion implantation technique can reduce friction by forming smoother surfaces by embedding the low-friction ions or materials, while laser surface treatment (LST) can reduce friction through surface texturing, creating micro-patterns that improve lubrication retention as well as minimising sliding friction between contacting surfaces [1]. Adhesion or bonding with coatings, adhesives, or paints can be improved by introducing functional groups through plasma surface treatment (or plasma activation) [2]. The introduction of specific ions on the surface of the material can also enhance the corrosion resistance by forming stable passivating layers. For instance, plasma nitriding treatment was applied to Ti6Al4V alloy, fabricated by electron beam melting (EBM), to enhance corrosion resistance by introducing a hard nitride compound surface layer that protected the alloy against oxidation and corrosion [3]. The electron beam surface treatment (EBST) technique can increase wear resistance through grain refinement and the formation of harder surface layers via rapid heating and cooling [4]. For example, the hardness, wear, and corrosion properties of 316 L austenitic stainless steel (ASS) were enhanced by EBST through microstructural refinement of the surface and subsurface of the material [5].

The ASM techniques such as laser treatment, plasma treatment, ion implantation, and electron beam treatment have thus been developed to enhance the tribological properties of components used in these industries. The development and production of high-performance materials are made possible by these techniques, which provide sophisticated control over the surface properties. This is critical in enhancing the service life of components, reducing operational costs, reducing environmental impacts, and contributing to the overall advancement of engineering technology. In this context, the advanced technologies employed for surface modification of industrial components, providing in-depth insights into their principles and applications, are explored. This chapter includes a review of the state-of-the-art and industrial applications of these ASM techniques, which enable the development of novel materials that meet the stringent demands of modern engineering applications. The objectives of this chapter are thus to:

- Understand the fundamental principles governing the various ASM techniques.
- Understanding how each technique improves material properties.
- Highlight the practical applications and benefits of these techniques in enhancing material properties such as hardness, corrosion resistance, and wear resistance.
- Discuss the emerging surface modification technologies, emphasising the potential innovations in the manufacturing sector.

2. Key principles of advanced surface modification

Recent studies highlight the importance of understanding the interactions between material's surface and its subsurface layers to achieve the desired modifications [6–8]. This is important in materials whose surfaces are usually exposed to severe operating conditions or high-stress environments, including thermal (low or high temperature exposure), wear, fatigue, and corrosion. Therefore, analyses of the chemical structures, as well as the mechanical and physical properties of the materials, are critical during the design, fabrication, and material testing stages. These analyses are necessary to successfully modify the surfaces of the materials and to ensure that the modified materials meet the required performance standards and functionality for their intended applications. The following examples underscore the importance of the analyses. During the design process, material selection and performance prediction are crucial, and there is numerous material design and simulation software to aid in this process. This is exemplified using nickel (Ni)-based superalloys for aircraft turbine blades due to their high-temperature resistance, low thermal expansion, and good fatigue resistance, necessitating an understanding of crystal structure and phase stability at high temperatures to predict performance and longevity [9]. Process optimisation and quality control are essential in the fabrication process. For instance, analysing the microstructure of the clad layer in laser cladding of steel components, such as high-carbon steel laser-clad with cobalt (Co)-based alloy to enhance wear resistance, requires understanding the alloy's phase formation and cooling rates to achieve a uniform microstructure with minimal defects, such as porosity and cracks [10]. During the testing stage, performance validation and failure analysis are critical. For example, titanium (Ti) alloys surface-modified with coatings for improved biocompatibility and osseointegration require testing mechanical properties like tensile strength and fatigue resistance, as well as biological properties such as cell attachment and growth, to ensure the success of implants in the human body [11]. This may require the use of machine learning to design new alloy compositions and predict their macroscale properties to a given design criterion for modern materials fabrication.

3. Overview of advanced surface modification techniques

3.1 Laser surface treatment

Laser surface treatment (LST) involves the irradiation of a high-energy laser beam directly at the substrate material or specimen's surface. Upon interaction with the substrate material, the high energy from the laser can heat, melt, or ablate the material, changing the surface topography, microstructure, and composition, and improving the material's surface properties [12]. The benefits of LST stem from the precise control the technique offers over the interaction with the material, allowing for targeted modifications. The laser-material interaction process is outlined in **Figure 1**. During the LST process, the excited electrons due to the laser beam collide with lattice ions and rapidly produce heat across a layer of the substrate, melting the layer and initiating complex phase transformations. As the laser moves, the molten layer cools rapidly, resulting in the formation of a new microstructure. The rapid cooling and solidification of the layer can introduce various phases, refine the grain structure, and consequently alter the chemical composition at the surface, leading to enhanced properties [12]. Specifically, LST can be used for enhancements such as surface hardening, where rapid heating and cooling form hard phases, increasing wear resistance and mechanical strength [13]. Additionally, the rapid cooling results in fine-grained microstructures that improve hardness and fracture toughness [14]. Laser surface

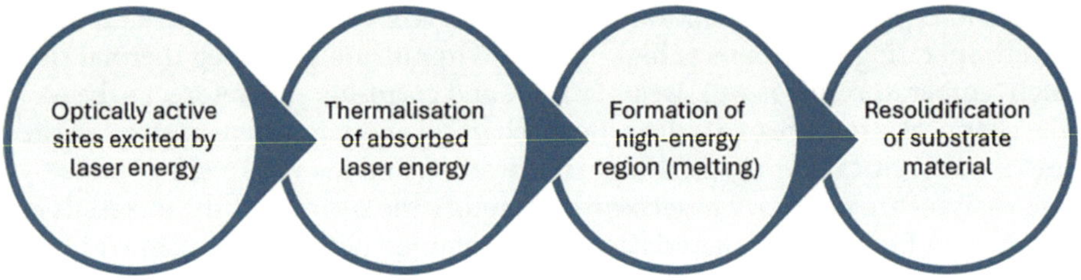

Figure 1.
Schematic showing the laser-material interactions.

treatment also enhances corrosion resistance by creating smoother surfaces with fewer defects, reducing corrosion initiation sites. Moreover, techniques like laser alloying and cladding allow for the introduction of new elements. For example, layers on the surface of magnesium (Mg) and its alloys using ceramics such as aluminium oxide (Al_2O_3) or silicon carbide (SiC) and metals such as Ni or cobalt (Co) are used to improve the hardness, mechanical strength, corrosion resistance, and wear resistance of Mg alloys [15]. The alloying layers can be formed by placing a suitable powder on

Technique (pulse duration)	Application	Material surface	Advantage	Disadvantage
Laser ablation (Nano- to femtosecond)	Precision micromachining, medical surgeries, material analysis	Metals, ceramic, polymers, composites	High precision (accuracy), minimal HAZ	Limited to thin material removal, higher equipment cost
Laser shock peening (Nanosecond)	Improving fatigue resistance, aerospace and automotive components	Metals (steel, aluminium, titanium)	Enhances material strength, minimal thermal effects	Requires special equipment
Laser cladding (Millisecond to continuous wave)	Adding wear or corrosion-resistant layers, and repairing and refurbishing worn-out parts	Metals	Improves surface properties, repairs worn parts	Significant HAZ, potential for thermal distortion
Laser texturing (Nano- to femtosecond)	Creating surface patterns, enhancing adhesion, reducing friction	Metals, polymers, ceramics	Versatile in creating various textures, minimal HAZ	Can be slower, requires precise control
Laser surface hardening (Millisecond)	Increasing wear resistance and hardness of metal surfaces, automotive, and aerospace components (i.e., gears and bearings)	Metals	Improves surface hardness, relatively simple process	Significant HAZ
Laser alloying (Millisecond to continuous wave)	Improving surface properties by adding alloying elements	Metals	Enhances surface composition and properties	Significant HAZ, potential for thermal distortion

Table 1.
Various laser techniques and their applications.

the surface, followed by laser melting and cooling [15]. **Table 1** shows the different laser surface treatment techniques and their respective applications [6].

Figure 2 shows the schematic diagram of a typical LST process. Several parameters, including the laser's energy density, wavelength, scanning speed, pulse duration, process environment, and the material's absorption properties, govern the interactions between the laser and the substrate material [16]. Metals, alloys, ceramics, or polymers have different energy absorption characteristics that affect their interaction with the laser beam. For instance, metals generally have high reflectivity, while polymers, in contrast, tend to have lower reflectivity. As a result, the energy absorption coefficient of these materials would vary for different processing wavelengths [7]. Understanding these interactions is crucial for optimising the laser treatment process to avoid introducing defects into the material. For instance, the high temperature gradient caused by the long-pulsed laser energy can lead to a pronounced heat-affected zone (HAZ). Hence, minimising the HAZ is critical to avoid compromising the bulk material properties [17]. Close control of the laser parameters is thus necessary to obtain appropriate surface properties for specific applications. Various techniques are employed in LST, including laser alloying, laser ablation, laser shock peening, laser cladding, laser texturing, and laser surface hardening, each with distinct mechanisms and applications.

3.2 Plasma surface treatment

Plasma surface treatment (PST) involves the generation of plasma—an ionised gas containing electrons, ions, and neurons under low vacuum, atmospheric pressure, or low temperature conditions to modify the surface properties of materials [18]. This treatment enhances surface properties through processes such as etching, activation, cleaning, and deposition. When the high-energy plasma interacts with the material's surface, it induces both physical and chemical modifications. These modifications

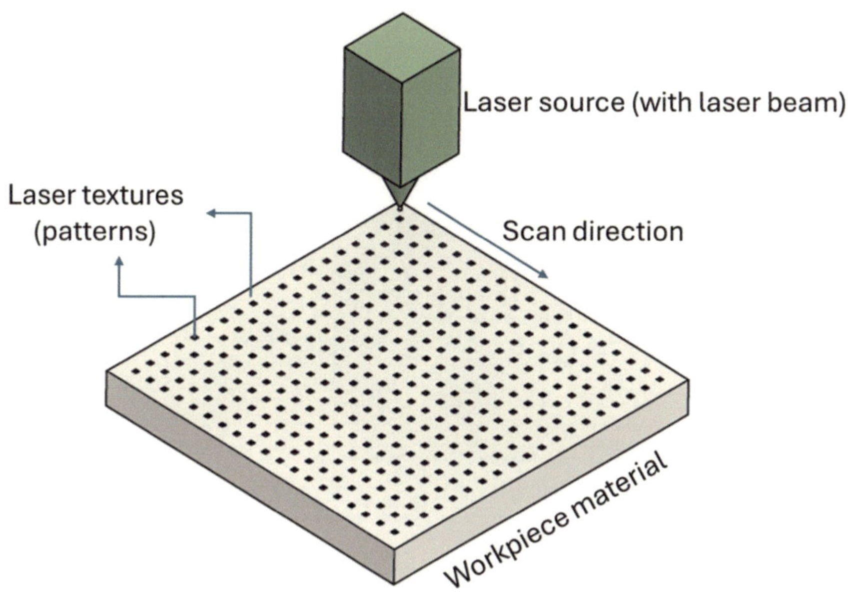

Figure 2.
Schematic illustration of the laser surface treatment process.

can include the formation of inorganic layers, which significantly enhance surface properties such as hardness, wear and corrosion resistance, thermal stability, corrosion and wear resistance, and biocompatibility, thereby improving the material's overall durability and performance [19, 20]. For instance, PST can be used for producing functional coatings that can suppress the formation of oxides on the surface of semiconductors, leading to improved performance [21]. Recent advancements in PST have introduced several innovative techniques, each offering precise control over surface modifications to enhance material performance in various applications. For example, plasma nitriding diffuses nitrogen into a metal surface to form a hard nitride layer, significantly improving wear resistance, hardness, and fatigue life. Similarly, plasma carburising introduces carbon atoms into the surface, enhancing hardness and wear resistance, making it ideal for components under high mechanical stress [22]. Plasma-enhanced chemical vapour deposition (PECVD) is used to deposit thin, uniform coatings with excellent adhesion and tailored properties, essential for industries like semiconductors and solar cells. Plasma immersion ion implantation (PIII) modifies surfaces by implanting ions into the material surface, enhancing properties like hardness and corrosion resistance, especially for complex geometries [23]. Plasma electrolytic oxidation (PEO) creates thick, ceramic-like oxide coatings on metals, providing exceptional wear and corrosion resistance, making it valuable in aerospace, automotive, and biomedical applications [22, 24]. It is worth noting that PST is one of the most versatile ASM techniques. Hence, the technique has been widely used in

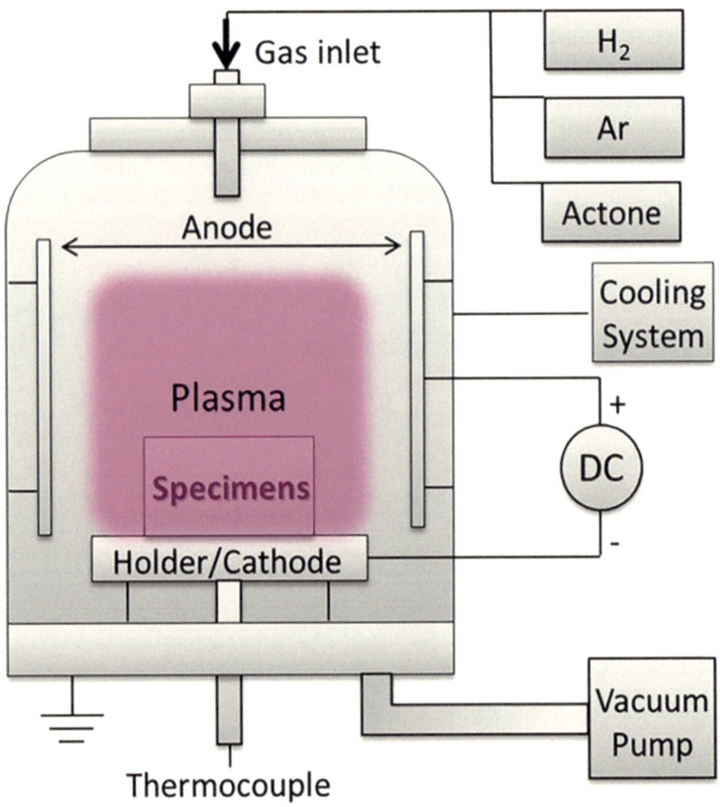

Figure 3.
A schematic diagram showing the plasma surface treatment [18].

automotive, aerospace [25–27], and biomedical [28] applications. However, despite the versatility and extensive advantages of PST, there are limitations to widespread adoption relating to high operational cost and precise control of process parameters. Moreover, the plasma treatment depth is usually limited to a few micrometres (μm) which may not be sufficient for applications requiring deeper modifications. **Figure 3** illustrates a schematic of a typical plasma surface treatment process [29]. Key parameters that must be controlled for achieving desired surface modifications when performing PST are given in **Table 2** [19, 30, 31].

Recent advancements in plasma surface treatment have introduced several innovative techniques, each offering precise control over surface modifications to enhance material performance in various applications. For example, plasma nitriding diffuses nitrogen into a metal surface to form a hard nitride layer, significantly improving wear resistance, hardness, and fatigue life. Similarly, plasma carburising introduces carbon atoms into the surface, enhancing hardness and wear resistance, making it ideal for components under high mechanical stress [22]. Plasma-enhanced chemical vapour deposition (PECVD) is used to deposit thin, uniform coatings with excellent adhesion and tailored properties, essential for industries like semiconductors and solar cells. Plasma immersion ion implantation (PIII) modifies surfaces by implanting ions into the material surface, enhancing properties like hardness and corrosion

Parameter	Effect
Plasma gas type	Determines the chemical reactions at the surface and influences the surface composition and functional groups. Common gases used in PST are grouped into different categories as follows [19]: • Oxidising gases—O_2, N_2O, H_2O, and air • Reducing gases—H_2 • Nitrogen and nitrogen-containing gases—N_2 and NH_3 • Fluorine containing gases—CF_4, and SF_6 • Carbon containing gases—CH_4 and C_2H_6 • Polymerising gases
Power input and plasma frequency	Controls the energy and behaviour of plasma species, affecting the intensity, efficiency, and depth of treatment. The different types of plasma excitation frequency include [19]: • Direct current (DC) and low-frequency discharges (i.e., corona and dielectric barrier discharges) • Radiofrequency • Microwave
Pressure and plasma density	Influences plasma characteristics, including particle mean free path, which affects the distribution and effectiveness of the treatment.
Exposure time and substrate temperature	Determines the extent of surface modification and reaction kinetics. Londer durations of exposure may lead to significant changes but may also increase the risk of overtreatment. Ref [17] demonstrated that increasing the plasma treatment time from 3 to 8 minutes increased the surface roughness, indicating potential interface damage.
Electrode configuration and distance	Affects plasma distribution, treatment uniformity, and interaction intensity across the substrate surface.

Table 2.
Key process parameters for plasma surface treatment.

resistance, especially for complex geometries [23]. Plasma electrolytic oxidation (PEO) creates thick, ceramic-like oxide coatings on metals, providing exceptional wear and corrosion resistance, making it valuable in aerospace, automotive, and biomedical applications [22, 24]. Despite the versatility and extensive advantages of PST, there are limitations to widespread adoption relating to high operational cost and precise control of process parameters. Moreover, the plasma treatment depth is usually limited to a few micrometres (μm) which may not be sufficient for applications requiring deeper modifications.

3.3 Conventional ion implantation

Conventional ion implantation (CII), also known as ion beam processing, is a procedure that involves accelerating ions generated by an ion source in an electric field and bombarding them into the solid substrate surface, resulting in desirable modifications in the material's chemical, optical, and mechanical properties [23]. This technique differs with PIII significantly in process, equipment, and application. For example, CII uses a precisely directed ion beam in a vacuum to modify surface properties with high accuracy. This makes the CII technique ideal for applications such as semiconductor manufacturing where exact control over ion depth and distribution are crucial [32, 33]. However, CII is typically limited to treatment of non-complex parts of planar surfaces. In contrast, PIII immerses the material in a plasma and uses a pulsed high-voltage bias to attract ions uniformly to all exposed areas of the substrate, allowing it to treat complex and three-dimensional surfaces. While PIII is less precise in controlling ion depth and dosage, it is more versatile and cost-effective for large-scale applications involving irregular geometries, such as medical implants and aerospace components. Therefore, CII is best for applications requiring precision, whereas PIII excels in uniformly treating complex shapes. During CII treatment process, ion species, such as nitrogen, carbon, boron, or phosphorus, determine the chemical nature of the implanted layer and influence properties like conductivity, friction, corrosion resistance, or hardness [32]. For instance, CII of phosphorus (P) has been effectively used to improve the biocompatibility as well as wear and corrosion resistance of titanium (Ti)-based implants, thus prolonging the life of the implants inside the human body [34]. In the semiconductor industry, CII is critical for doping silicon wafers with impurities such as boron (B) of P to control electrical characteristics in transistors and diodes, which are essential for microchips and integrated circuits [35]. Cutting tools in manufacturing, such as drills and milling cutters, are also treated with this technique to form hard nitrides or carbides, boosting their longevity and cutting efficiency. For example, the tool life performance and wear resistance of WC-Co-based cutting tools were significantly improved by nitrogen ion implantation [36]. These diverse applications demonstrate the versatility of ion implantation in industries where improved material properties are key to operational success.

There are several key parameters that must be carefully controlled to achieve desired surface modification in the CII technique. The ion energy, measured in keV or MeV, controls the penetration depth of the ions into the material [23]. The dose (ion fluence) dictates the concentration of implanted ions and affects the surface's properties. The beam current impacts the rate of implantation, where higher currents speed up the implantation process but increase the risk of surface heating and damage. Lastly, the substrate temperature during the implantation affects how the material responds to the ions. Higher temperatures promote the diffusion and reduce lattice damage, while lower temperatures may preserve surface integrity but can result in more defects [37]. In a typical CII system, as shown in **Figure 4**, atoms of the desired species are ionised to create a plasma of ions. These ions are then directed through an orifice into a high-vacuum region. In this region, the ions are first accelerated by an electric field and then sorted according to their mass using a magnetised mass analyser. A set of extraction grids focuses the ions into a collimated beam. This finely focused beam is guided through a beam steering system, where the ions are accelerated to the desired energy level and then deflected across the target surface using an electrostatic scanning mechanism, ensuring uniform implantation. For nonplanar targets (components), a manipulator stage is employed to rotate the target, enabling implantation from all sides. This process is controlled by a computer system. In addition, the manipulator stage incorporates a heat sink at the target to prevent excessive temperature increase during implantation [23, 38].

3.4 Electron beam surface treatment

Electron beam surface treatment (EBST) is a process that utilises a concentrated high-energy beam of electrons to modify the surface of a material [39]. During the EBST processing, the raw material used is either metal powder or wire, which are fused to the surface of a solid material by an electron beam by melting [40]. This process leads to surface hardening and grain refinement through rapid heating and cooling, which enhances the surface properties including hardness, mechanical strength, wear, fatigue, corrosion resistance, and thermal stability. The intense energy from the electron beam can melt and re-solidify surface defects such as cracks, smoothing the surface, and increasing fatigue life. Electron beam surface treatment can also change the microstructure and chemical composition of the surface, creating denser, and oxidation-resistant layers that provide enhanced corrosion resistance and thermal stability, especially in high-temperature environments. This technique has recently been receiving a lot of attention and employed in a range of applications, including automotive, aerospace, military, tissue engineering, and medical [40]. The successful application of EBST in various industries is due to the possibility of accurate control of the processing conditions, short process time, homogeneous distribution of the electron beam energy, and ability to produce complex parts or treat surfaces with complex geometries [41].

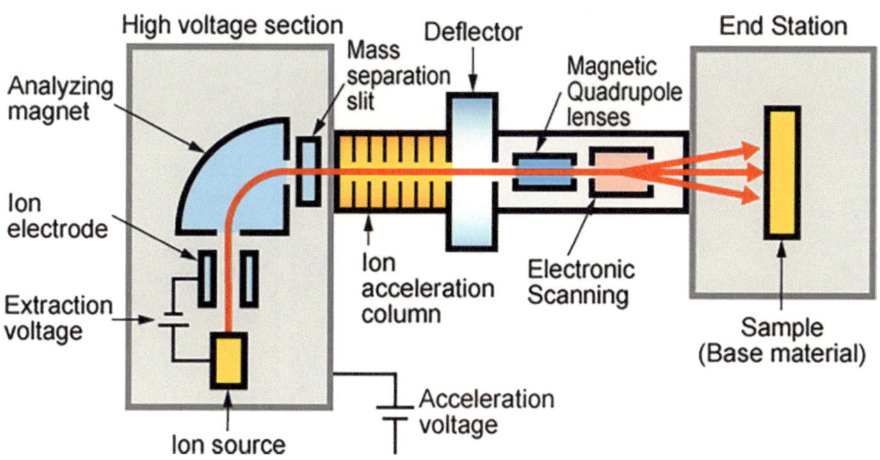

Figure 4.
Schematic diagram of a conventional (CII) process [28].

The process of EBST begins with a specialised system that includes a vacuum chamber, an electron source, which is typically a tungsten cathode, an electron gun, and magnetic lenses for beam focusing. **Figure 5** shows a schematic of a typical electron beam surface treatment facility [42]. The cathode emits electrons when heated, and these electrons are accelerated through a high-voltage field. The accelerated electrons are then focused into a narrow, intense beam by magnetic lenses and directed toward the material's surface within the vacuum chamber to prevent scattering and ensure precision. Upon impact with the surface, the kinetic energy of the electrons is converted into thermal energy, which causes localised heating, melting, or vaporisation of the material's surface. This precise heating allows for surface modifications such as hardening, alloying, or texturing. Like the other surface treatment techniques, several key parameters must be carefully controlled to ensure that desired surface modifications are achieved [39, 43]. The beam energy, typically measured in kV, determines the depth of electron penetration into the material, with higher energy resulting in deeper modification and lower energy affecting the surface only. The beam current regulates the intensity of the energy delivered to the surface, with higher currents increasing the heating and altering the surface more intensely. The beam focus or spot size controls the concentration of energy on the material. The scanning speed of the electron beam affects the duration of energy exposure, with

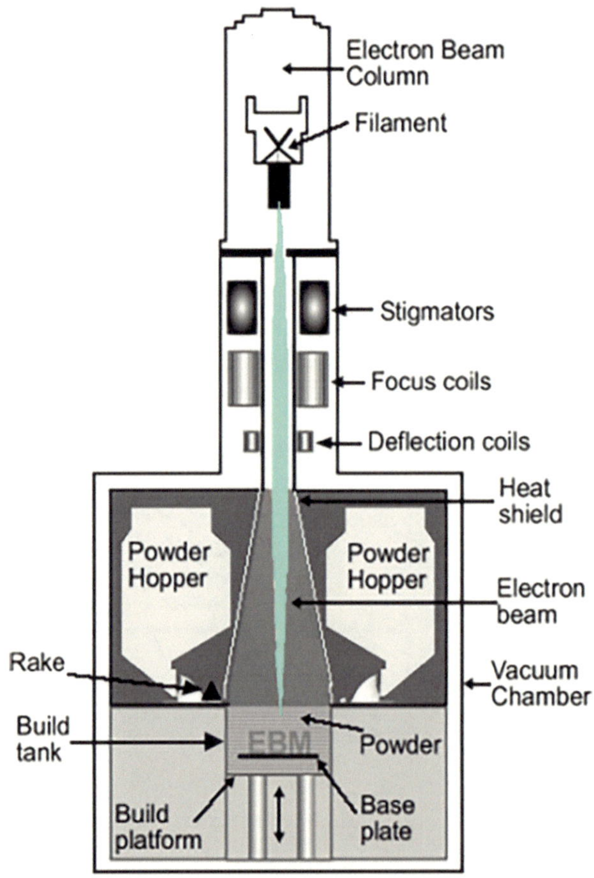

Figure 5.
Schematic of a typical electron beam surface melting system [31].

faster scanning leading to less intense surface heating and slower speeds resulting in deeper and more thorough surface modification [43]. Finally, the vacuum level in the processing chamber ensures that the electron beam precision remains focused and efficient by preventing electron scattering, which is essential for maintaining beam accuracy as well as achieving uniform surface treatment.

4. Case studies and real-world applications

Advanced surface modification techniques have revolutionised various industries by enhancing the performance and longevity of various materials under severe conditions. For instance, LST has been widely adopted in the automotive and aerospace industries to enhance the tribological properties of engine components, leading to increased efficiency and reduced maintenance costs [44, 45]. For example, Nagarajan et al. [46] explored femtosecond laser irradiation to selectively remove Co binder phase from the WC-Co surface, improving the adhesion between the WC-Co

Technique	Substrate	Treatment condition	Main effect	Ref.
LST	4H-SiC	Femtosecond laser irradiation	Fluctuating coefficient of friction, decrease in surface stiffness, increase in surface roughness	[49]
	Ti-10 V-2Fe-3Al	Nanosecond laser irradiation	Enhanced surface roughness, reduced coefficient of friction by 66% under dry conditions	[50]
	304 stainless-steel	Nanosecond laser irradiation	Tensile stresses dominant in surface and near-surface regions	[51]
	Ti-6Al-4 V	Femtosecond laser irradiation	Surface grain thinning, reduced surface wear (~74%)	[52]
PST	AISI 4140	Shot peened	Compressive residual stresses, lower wear rate	[53]
	Indium-tin-oxide film	O_2	Flat film surface morphology, stable surface sheet resistance	[54]
	Polymer	O_2 and N_2	Micro- or nanotexturing and generation of functional surface groups for facilitating linker-free immobilisation of biomolecules	[55]
CII	Ti6Al7Nb	N_2 immersion	Higher surface roughness, hardness, corrosion resistance in simulated body fluid	[20]
	Ti-25Nb-3Mo-2Sn-3Zr	Phosphorus ion implantation	Enhanced wear resistance, formation of TiP + β-Ti nanograins + P-interstitial amorphous hybrid layer, better surface bioactivity	[34]
EBST	Inconel 718	Vacuum	Enhanced corrosion resistance, wear resistance improved by 15%	[56]
	AlCoCrFeNi2.1		Slower fatigue crack growth rate	[57]
	Ti-6Al-4 V		Long curved martensite laths, increasing hardness	[58]

Table 3.
Applications of surface modification techniques across various industries.

substrate and a diamond coating. Comparatively, EBST has found significant applications in the medical field, particularly in the fabrication of implants to enhance surface hardness and corrosion resistance, thus ensuring biocompatibility and long-term durability [47, 48]. **Table 3** shows examples of real-world applications where surface modification techniques are employed. These case studies underpin the transformative impact of advanced surface modification techniques, demonstrating their critical role in advancing surface technology across diverse industries.

5. Conclusion

The fundamentals and applications of advanced surface modification techniques, including laser surface treatment (LST), plasma surface treatment (PST), conventional ion implantation (CII), and electron beam surface treatment (EBST), were discussed in this chapter. The successful application of these techniques is reliant on the understanding of processing parameters in relation to diverse materials. For instance, LST faces challenges related to process control and repeatability, especially when scaling up for large-area applications in the automotive and aerospace industries. Precise LST process control is required to achieve uniform surface modification on complex geometries [59]. Comparatively, EBST requires the use of a vacuum chamber, which tends to increase the equipment price, thus limiting its accessibility for widespread industrial adoption. This is critical in the medical industry, where precision and biocompatibility are important. Conventional ion implantation, on the other hand, is plagued by challenges in depth control without altering the bulk material properties. These challenges highlight the need for further research to address the process limitations of the techniques, as well as concerns around resource management and climate change. In line with this, the ASM research outlook of the aerospace, automotive, and medical industries is summarised as follows:

- *Aerospace*: ASM in aerospace should be centred around the development of high-strength and ultra-lightweight components that can withstand harsh conditions. LST and EBST processes can be explored to enhance wear resistance and thermal stability of critical parts such as the turbine blades and the fuselage components. These techniques may be most beneficial for the surface treatment of additively manufactured components.

- *Automotive*: For the automotive industry, research should focus on the development of surface treatments that can improve fuel efficiency, optimise performance, and reduce carbon emissions. With the shift toward electrical cars, there will be a growing demand for advanced surface treatments to enhance the performance and durability of lightweight materials and batteries. Techniques such as plasma nitriding and PECVD coatings can reduce friction and wear in engine, transmission, and braking components, while LST and EBST can enhance the lifespan of lightweight alloys and improve the efficiency of battery components. These advancements will be essential in boosting the performance, longevity, and sustainability of next-generation electric vehicles.

- *Medical*: As personalised medicine and minimally invasive medical techniques evolve, the development of biocompatible and antibacterial surfaces for implants and medical devices should be the focus of medical surface modification research. This may include exploring novel coatings and modification techniques. Moreover, advances in nanotechnology and ion implantation are going to play a critical role in medical research, enabling the development of wear- and corrosion-resistant devices that are tailored for individual medical needs.

In addition to established techniques discussed in this chapter, several emerging and cutting-edge surface modification techniques such as atomic layer deposition (ALD), cold spray (CS), magnetron sputtering (MS), femtosecond laser surface structuring, bio-inspired coatings, and nanostructured coatings are gaining prominence and revolutionising various industries. For instance, ALD allows for atomic-level control of coating thickness and composition, ensuring precise thin-film deposition [60]. Cold spray is a solid-state process that enhances wear and corrosion resistance without thermal damage, ideal for aerospace and additive manufacturing [61]. Magnetron sputtering offers efficient deposition of dense, durable coatings, commonly used in semiconductor and hard tool industries [62]. Femtosecond layer surface structuring provides high precision patterning (texturing) for creating micro- and nanostructures in various industries [63]. Bio-inspired coatings offer superior properties such as self-cleaning, enhanced adhesion, and anti-fouling capabilities [64], suitable for advanced fabrication of medical devices, sensors, and energy systems owing to their highly functional and environmentally friendly capabilities. Nanostructured coatings leverage nano-sized grains for improved physical, chemical, and mechanical properties for various applications [65].

Overall, the prospects of surface modification techniques across various industries are characterised by a strong emphasis on better control over processing and surface properties, scalability, sustainability, data-driven decision-making, and broader material compatibility with a wide range of materials, including additively manufactured components. However, several challenges, such as process efficiency, scalability, and cost-effectiveness, exist that could impact the development and widespread adoption of both established and emerging surface modification techniques. Scalability continues to be a significant difficulty for established techniques like LST, PST, CII, and EBST. While these techniques improve surface qualities, including wear resistance, adhesion, and corrosion protection, the necessary equipment and processes can be expensive and time-consuming for large-scale industrial applications. For example, laser and ion implantation procedures frequently need precise control, limiting their throughput in high-volume production environments and posing substantial challenges to scalability and cost-effectiveness. Technical challenges also pose challenges for both established and emerging techniques. For example, PST often struggles with homogeneous coating over complex geometries, while EBST may suffer from issues such as limited depth of penetration, restricting their application to certain material types and thicknesses. Similarly, for emerging techniques like ALD and CS, a key challenge is ensuring consistency and control over the deposition process parameters, especially when dealing with complex shapes or nanoscale features. Atomic layer deposition, though offering atomic-level control, is typically time-consuming and can be difficult for large-scale production or complex component applications, limiting its scalability. For CS, challenges such as achieving high bond strength and uniform coating thickness in different materials require further refinement. Cost is another critical challenge for both advanced and emerging surface modification techniques. Many of these processes involve expensive equipment and energy-intensive operations, which could limit their adoption in industries with tighter profit margins or where surface modification is only a small part of the overall manufacturing process. For instance, femtosecond laser surface structuring offers excellent precision but at a high operational cost, which makes it less desirable for industries where cost-effectiveness is paramount. Finally, sustainability is becoming a more essential consideration in the development of ASM techniques. Some processes involve toxic chemicals or require a lot of energy, which raises concerns about their environmental impact. Emerging approaches such as nanostructured coatings might offer more sustainable alternatives, but their adoption is dependent on overcoming material compatibility and cost barriers. Data-driven decision-making and process optimisation using machine learning and artificial intelligence could help address some of these issues by increasing efficiency, decreasing waste, and improving control over surface qualities. However, integrating these technologies necessitates substantial investment and technical skill, adding another degree of complexity. In summary, ASM techniques continue to face scalability, cost, technical limits, and sustainability issues that must be addressed through continuous research and development for industry-wide adoption.

Conflict of interest

The authors declare that they have no conflict of interest in the work.

Author details

Nthape P. Mphasha[1*] and Mahlatse S. Rabothata[2]

1 School of Chemical and Metallurgical Engineering, University of the Witwatersrand, Johannesburg, South Africa

2 Academic Development Unit, University of the Witwatersrand, Johannesburg, South Africa

*Address all correspondence to: nthape.mphasha@wits.ac.za

References

[1] Wang Z, Ye R, Xiang J. The performance of textured surface in friction reducing: A review. Tribology International. 2023;**177**:108010. DOI: 10.1016/J.TRIBOINT.2022.108010

[2] Scarselli G, Quan D, Murphy N, Deegan B, Dowling D, Ivankovic A. Adhesion improvement of thermoplastics-based composites by atmospheric plasma and UV treatments. Applied Composite Materials. 2021;**28**(1):71-89. DOI: 10.1007/s10443-020-09854-y

[3] Li Y et al. Improved seawater corrosion resistance of electron beam melting Ti6Al4V titanium alloy by plasma nitriding. Vacuum. 2023;**216**:112463. DOI: 10.1016/j.vacuum.2023.112463

[4] Ivanov YF et al. Changes in surface structure and mechanical characteristics of Al–5 wt%Si alloy after irradiation by electron beam. Materials Letters. 2020;**275**:128105. DOI: 10.1016/J.MATLET.2020.128105

[5] Basak S et al. Electron beam surface treatment of 316L austenitic stainless steel: Improvements in hardness, wear, and corrosion resistance. Metals and Materials International. 2021;**27**(5):953-961. DOI: 10.1007/s12540-020-00773-y

[6] El-Awadi GA. Review of effective techniques for surface engineering material modification for a variety of applications. Aims Material Science. 2023;**10**(4):652-692. DOI: 10.3934/MATERSCI.2023037

[7] Ravi-Kumar S, Lies B, Lyu H, Qin H. Laser ablation of polymers: A review. Procedia Manufacturing. 2019;**34**:316-327. DOI: 10.1016/J.PROMFG.2019.06.155

[8] Valkov S, Dechev D, Ivanov N, Bezdushnyi R, Ormanova M, Petrov P. Influence of beam power on young's modulus and friction coefficient of Ti–Ta alloys formed by electron-beam surface alloying. Metals (Basel). 2021;**11**(8):1246. DOI: 10.3390/met11081246

[9] Gudivada G, Pandey AK. Recent developments in nickel-based superalloys for gas turbine applications: Review. Journal of Alloys and Compounds. 2023;**963**:171128. DOI: 10.1016/J.JALLCOM.2023.171128

[10] Yao J, Zhang Q, Liu R, Wu G. Laser cladding of satellite alloys. In: Yao J, Zhang Q, Liu R, Wu G, editors. Laser Applications in Surface Modification. Singapore: Springer Singapore; 2022. pp. 11-57. DOI: 10.1007/978-981-16-8922-2_2

[11] Kumar A, Singh G. Surface modification of Ti6Al4V alloy via advanced coatings: Mechanical, tribological, corrosion, wetting, and biocompatibility studies. Journal of Alloys and Compounds. 2024;**989**:174418. DOI: 10.1016/J.JALLCOM.2024.174418

[12] Jeyaprakash N, Yang C-H, Raj Kumar D. Laser surface modification of materials. In: Practical Applications of Laser Ablation. London, UK: IntechOpen; 2021. DOI: 10.5772/intechopen.94439

[13] Wang B et al. Effects of quench-tempering and laser hardening treatment on wear resistance of gray cast iron. Journal of Materials Research and Technology. 2020;**9**(4):8163-8171. DOI: 10.1016/J.JMRT.2020.05.006

[14] Genga RM, et al. Laser shock peening: A NbC based cermet

enhancement alternative for improved GCI interrupted face-milling. In: von Leipzig,K, Sacks N, Mc Clelland M, editors. Smart, Sustainable Manufacturing in an Ever-Changing World. Lecture Notes in Production Engineering. Cham: Springer; 2023. DOI: 10.1007/978-3-031-15602-1_62

[15] Kumar PS et al. State of art: Review on laser surface hardening of alloy metals. Materials Today Proceedings. 2023. DOI: 10.1016/J.MATPR.2023.04.259 (In Press)

[16] Denkena B, Krödel A, Ellersiek L, Murrenhoff M. Production of chip breakers on cemented carbide tools using laser ablation. Procedia CIRP. 2020;**94**:834-839. DOI: 10.1016/J.PROCIR.2020.09.114

[17] Zhou R, Zhang Z, Hong M. The art of laser ablation in aeroengine: The crown jewel of modern industry. Journal of Applied Physics. 2020;**127**(8):080902. DOI: 10.1063/1.5134813

[18] Kaseem M, Fatimah S, Nashrah N, Ko YG. Recent progress in surface modification of metals coated by plasma electrolytic oxidation: Principle, structure, and performance. Progress in Materials Science. 2021;**117**:100735. DOI: 10.1016/J.PMATSCI.2020.100735

[19] Sikdar S, Menezes PV, Maccione R, Jacob T, Menezes PL. Plasma Electrolytic Oxidation (PEO) Process-Processing, Properties, and Applications. Materials. 2021;**11**(6):1375. DOI: 10.3390/nano11061375

[20] Huang HH, Liu CF, Wang S, Chen CS, Chang JH. Nitrogen plasma immersion ion implantation treatment of Ti6Al7Nb alloy for bone-implant applications: Enhanced in vitro biological responses and in vivo initial bone-implant contact. Surface and Coating Technology. 2021;**405**:126551. DOI: 10.1016/J.SURFCOAT.2020.126551

[21] Xu Q, Liu WD, Ding YX, Zheng ZW, Ying LY, Zhang BP. Effects of fluorine plasma pre-treatment on electrical properties of high-κ-based InP metal-oxide-semiconductor device. Applied Surface Science. 2022;**585**:152688. DOI: 10.1016/j.apsusc.2022.152688

[22] Lamim TS, Bernardelli EA, Bendo T, Mello CH, Binder C, Klein AN. Duplex surface treatment of sintered iron by plasma nitriding and plasma carburizing at low temperature. Surface and Coating Technology. 2019;**375**:911-919. DOI: 10.1016/j.surfcoat.2019.07.068

[23] Gupta D. Plasma immersion ion implantation (PIII) process-physics and technology. International Journal of Advancements in Technology. 2011;**2**(4):471. Available from: http://ijict.org/

[24] Akdoğan E, Şirin HT. Plasma surface modification strategies for the preparation of antibacterial biomaterials: A review of the recent literature. Materials Science and Engineering: C. 2021;**131**:112474. DOI: 10.1016/j.msec.2021.112474

[25] dos Santos Filho D, Tschiptschin AP, Goldenstein H. Effects of shallow plasma nitriding on the surface topography of gray cast iron specimens. Surface and Coating Technology. 2020;**404**:126464. DOI: 10.1016/j.surfcoat.2020.126464

[26] Shin DH, Hwang HK, Kim SJ. Effect of plasma ion nitriding on hydrogen embrittlement and wear resistance of martensitic stainless steel for hydrogen valve core of fuel cell electric vehicles. Engineering Fracture Mechanics. 2024;**306**:110247. DOI: 10.1016/j.engfracmech.2024.110247

[27] Zhang H et al. Surface melt nitriding and strengthening mechanism of plasma torched M50 bearing steel. Journal of Manufacturing Processes. 2024;**124**:1214-1226. DOI: 10.1016/j.jmapro.2024.07.012

[28] Chou WC, Wang RCC, Huang CL, Lee TM. The effect of plasma treatment on the osseointegration of rough titanium implant: A histo-morphometric study in rabbits. Journal of Dental Science. 2018;**13**(3):267-273. DOI: 10.1016/j.jds.2018.06.002

[29] Yang Y, Yan MF, Zhang SD, Zhang YX, Chen HT, Wang XA. One-step plasma-assisted method for functionally graded Fe_3O_4/DLC coated carburized layer on steel. Diamond and Related Materials. 2016;**70**:18-25. DOI: 10.1016/j.diamond.2016.09.020

[30] Sasmazel HT, Alazzawi M, Alsahib NKA. Atmospheric pressure plasma surface treatment of polymers and influence on cell cultivation. Molecules. 2021;**26**(6):1665. DOI: 10.3390/molecules26061665

[31] Dufour T. From basics to frontiers: A comprehensive review of plasma-modified and plasma-synthesized polymer films. Polymers. 2023;**15**(17):3607. DOI: 10.3390/polym15173607

[32] Ren X et al. Surface modification technologies for enhancing the tribological properties of cemented carbides: A Review. Tribology International. 2023;**180**:108257. DOI: 10.1016/j.triboint.2023.108257

[33] Wood J, Majumdar G. Ion implantation. In: Reference Module in Materials Science and Materials Engineering. Amsterdam, Netherlands: Elsevier; 2016. DOI: 10.1016/B978-0-12-803581-8.03724-3

[34] Huang R et al. Enhancing wear resistance, anti-corrosion property and surface bioactivity of a beta-titanium alloy by adopting surface mechanical attrition treatment followed by phosphorus ion implantation. Applied Surface Science. 2024;**671**:160766. DOI: 10.1016/j.apsusc.2024.160766

[35] Current MI. Ion implantation of advanced silicon devices: Past, Present and Future. Materials Science in Semiconductor Processing. 2016;**16**:13-22. DOI: 10.1016/j.mssp.2016.10.045

[36] Wilkowski J et al. Effect of nitrogen ion implantation on the lifetime of WC-Co tools used in particleboard milling. Wood Material Science & Engineering. 2022;**17**(6):521-532. DOI: 10.1080/17480272.2021.1900391

[37] Qing Li W et al. The Ion Implantation-Induced Properties of One-Dimensional Nanomaterials. 2013. Available from: http://www.nanoscalereslett.com/content/8/1/175

[38] Wang X et al. Application of ion beam technology in (photo) electrocatalytic materials for renewable energy. Applied Physics Reviews. 2020;7:041303. DOI: 10.1063/5.0021322

[39] Valkov S, Ormanova M, Petrov P. Electron-beam surface treatment of metals and alloys: Techniques and trends. Metals (Basel). 2020;**10**(9):1-20. DOI: 10.3390/met10091219

[40] Sonkamble V, Phafat N. A current review on electron beam assisted additive manufacturing technology: Recent trends and advances in materials design. Discover Mechanical Engineering. 2023;**2**(1). DOI: 10.1007/s44245-022-00008-x

[41] Valkov S, Ormanova M, Petrov P. Surface manufacturing of materials

by high energy fluxes. In: Advanced Surface Engineering Research. London, UK: InTech; 2018. DOI: 10.5772/intechopen.79874

[42] Sidambe AT. Biocompatibility of advanced manufactured titanium implants - A Review. Materials. 2014;**7**(12):8168-8188. DOI: 10.3390/ma7128168

[43] Grafe S, Hengst P, Buchwalder A, Zenker R. Investigations on the influence of parameters during electron beam surface hardening using the flash technique. In: IOP Conference Series: Materials Science and Engineering, Institute of Physics Publishing. Chemnitz, Germany: IOP Publishing Ltd; 2018. DOI: 10.1088/1757-899X/373/1/012019

[44] Bao Y, Deng J, Cao S, Ma K, Zhang Z, Lu Y. Laser micro-cladding in situ forming textured surface to improve the tribological performance. Wear. 2024;**550-551**:205422. DOI: 10.1016/j.wear.2024.205422

[45] Saetang V, Taweejun N, Zhu H, Jaafar H, Daodon W. Effects of laser texturing on tribological properties and wettability of titanium nitride-coated AISI M2 high-speed steel. Surface and Coating Technology. 2024;**490**:131173. DOI: 10.1016/j.surfcoat.2024.131173

[46] Nagarajan B, Han J, Huang S, Qian J, Vleugels J, Castagne S. Femtosecond laser processing of cemented carbide for selective removal of cobalt. Procedia CIRP. 2022;**113**:576-581. DOI: 10.1016/J.PROCIR.2022.09.177

[47] Gotzmann G, Beckmann J, Wetzel C, Scholz B, Herrmann U, Neunzehn J. Electron-beam modification of DLC coatings for biomedical applications. Surface and Coating Technology. 2017;**311**:248-256. DOI: 10.1016/j.surfcoat.2016.12.080

[48] Sheng K, Dong X, Chen Z, Zhou Z, Gu Y, Huang J. Increasing the surface hydrophobicity of silicone rubber by electron beam irradiation in the presence of a glycerol layer. Applied Surface Science. 2022;**591**:153097. DOI: 10.1016/j.apsusc.2022.153097

[49] Huang Y, Zhou Y, Li J, Zhu F. Femtosecond laser surface modification of 4H-SiC improves machinability. Applied Surface Science. 2023;**615**:156436. DOI: 10.1016/j.apsusc.2023.156436

[50] Guo J, Hu T, Li Q, Liu Y. Nanosecond laser-induced dimple texturing of TB6 alloy surfaces: Tribological behavior under dry and starved-oil lubrication. Tribology International. 2024;**197**:109842. DOI: 10.1016/j.triboint.2024.109842

[51] Zhao K, Pan J, Chen X. Texture morphologies and residual stresses due to laser surface texturing on the 304 stainless steels. Applied Thermal Engineering. 2024;**255**:123951. DOI: 10.1016/j.applthermaleng.2024.123951

[52] Li LX, Li ZX, Xing ZG, Guo WL, Huang YF, Wang HD. Effect of femtosecond laser bionic texture on anti-wear properties of medical Ti-6Al-4V. Tribology International. 2023;**190**:109062. DOI: 10.1016/j.triboint.2023.109062

[53] Çelik H, Kovacı H. Combined effects of magnetic field and surface treatments consisting of shot peening, plasma nitriding and their consecutive application on the friction and wear behavior of a ferromagnetic steel. Materials Today Communications. 2023;**37**:107467. DOI: 10.1016/j.mtcomm.2023.107467

[54] Chen PH et al. Effects of plasma treatment time on surface characteristics

of indium-tin-oxide film for resistive switching storage applications. Applied Surface Science. 2017;**414**:224-229. DOI: 10.1016/j.apsusc.2017.04.060

[55] Wieland F, Bruch R, Bergmann M, Partel S, Urban GA, Dincer C. Enhanced protein immobilization on polymers-a plasma surface activation study. Polymers. 2020;**12**(1):104. DOI: 10.3390/POLYM12010104

[56] Zhao G, Zhang P, Li J, Zhang Z, Li H, Ma L. Effects of different scanning speeds on microstructure evolution and tribological properties of Inconel 718 alloy vacuum electron beam surface modification. Materials Today Sustainability. 2024;**25**:100613. DOI: 10.1016/j.mtsust.2023.100613

[57] Zheng W et al. Microstructure evolution and fatigue crack propagation of AlCoCrFeNi2.1 high entropy alloy after electron beam surface modification. Materials Science and Engineering: A. 2024;**913**:147025. DOI: 10.1016/j.msea.2024.147025

[58] Yan X, Xu X, Zhou Y, Wu Z, Wei L, Zhang D. Surface electropulsing-induced texture evolution in electron beam melted Ti-6Al-4V alloy for biomedical application. Surface and Coating Technology. 2024;**479**:130509. DOI: 10.1016/j.surfcoat.2024.130509

[59] Zhang B, Sun L, Zhao N, Li J. Applications of laser surface treatment in gears: A review. Journal of Materials Engineering and Performance. 2024;**33**:DOI: 10.1007/s11665-024-09945-y

[60] Oke JA, Jen TC. Atomic layer deposition and other thin film deposition techniques: From principles to film properties. Journal of Materials Research and Technology. 2022;**21**:2481-2514. DOI: 10.1016/J.JMRT.2022.10.064

[61] Dai S, Cui M, Li J, Zhang M. Cold spray technology and its application in the manufacturing of metal matrix composite materials with carbon-based reinforcements. Coatings. 2024;**14**(7):822. DOI: 10.3390/coatings14070822

[62] An Q, Jiang T, Fang H. Improvement of titanium film uniformity by magnetron sputtering with electromagnetic coil design. Surface and Coating Technology. 2024;**494**:131365. DOI: 10.1016/J.SURFCOAT.2024.131365

[63] Genga RM et al. Roughing, semi-finishing and finishing of laser surface modified nickel bonded NbC and WC inserts for grey cast iron (GCI) face-milling. International Journal of Refractory Metals and Hard Materials. 2019;**86**:105128. DOI: 10.1016/j.ijrmhm.2019.105128

[64] Ni X et al. Design of chlorella-inspired and smart-responsive bio-microspheres and their application in eco-friendly anti-biofouling coatings. Chemical Engineering Journal. 2023;**478**:147426. DOI: 10.1016/J.CEJ.2023.147426

[65] duo Ma Y, Yang Y, Wu Yuan L, Tian W, Zhao H. Reinforcing and toughening mechanism of the in-situ metastable nanostructured alumina-titanium oxide composite coating. Journal of Materials Science and Technology. 2025;**208**:120-131. DOI: 10.1016/J.JMST.2024.04.060

Chapter 3

Powertrain Tribology Development Trends and Impact of Surface Engineering on Friction and Wear Control

Simon C. Tung

Abstract

In this chapter, the fundamentals of surface coatings or textures specific to the environments of powertrain components will be reviewed, together with discussions on the impact of surface engineering on friction and wear. In addition, the advanced surface coating methods will be described, based on promises of continuing friction and wear reduction trends. Specifically, this paper will address the impact of these emerging technologies on future powertrain performance requirements using novel surface coatings or texture materials. The connection between surface engineering and tribological performance requirements will be illustrated by briefly describing the surface engineering processes incorporating the emerging powertrain technologies. Surface engineering technology using coatings or textures has successfully applied to advanced powertrain components fabricated from non-ferrous lightweight materials to meet required tribological performance, energy efficiency, and fuel economy demands. Lastly, besides new hardware and material science changes, all advanced surface coatings or textures represent tribological solutions to meet more stringent energy efficiency and durability requirements used in powertrain components. The final section of this chapter will review and discuss the future development trends of non-ferrous lightweight materials in powertrain components by surface coatings or textures for friction and wear control.

Keywords: automotive tribology, surface engineering, friction, wear control, surface coatings

1. Introduction

Recently surface engineering technologies have been successfully applied to improve engineering material performance across automotive and manufacturing industries. Light-weight non-ferrous alloy and composite materials are applied to powertrain or drivetrain components to improve energy efficiency, fuel economy, and wear resistance. For example, the thermally sprayed aluminum engines, chemical vapor deposited coated bearings, and diamond-like-carbon coated piston rings used

IntechOpen

in valve train and transmission components [1, 2]. Surface coatings or textures also enable a variety of engineering materials to be used in powertrain components, which can provide tribological performance and long-term durability benefits.

This chapter will provide a comprehensive overview of the latest developments and future trends of friction reduction and wear control by applications of surface coatings or textures in powertrain or drivetrain components. Industrial surface coatings or modified surface textures have been widely applied to powertrain components for friction control and wear reduction. Current automotive industries have developed more surface coatings applications for powertrain and driveline components such as engine cylinder bores, bearings, engine pistons, piston rings, driveline transmission, and gear components. Surface coatings in automotive industries have been applied for improved energy efficiency and wear life of powertrain components [3, 4]. The future development trends of powertrain components will be extended by using advanced lubricant additive formulation under severe operating conditions such as extreme temperatures or heavy loading applications. The lubricant additives reacted with the coated metallic surfaces can generate effective tribochemical film formation for friction and wear control.

In this review chapter, tribological characteristics of surface coatings or textures can be influenced by their operating parameters such as the contact geometries, oil temperature, sliding speed, and the applied load. In addition, automotive engineers and lubricant formulation scientists have applied novel surface engineering methods to protect the powertrain components from wear, keeping the powertrain system working at the optimal fuel economy while retaining long-term durability.

To reduce friction and wear in powertrain components, automotive suppliers have applied the following approaches: (1) mechanical design by the improved micro-geometries, configurations, and properties of the major components These major powertrain components such as the journal bearings, crankshaft bearings, camshaft bearings, and valve-train components dominate the large portion of the frictional losses. These valve-train and bearing systems do have different tribological characteristics which depend on the specific design of the engine [3, 4]. Boundary lubrication dominates the valve-train system where surface contact stress is extremely high. However, the major lubrication modes in the crankshaft and camshaft bearings are hydrodynamic lubrication. Different lubrication modes have influenced the lubricant and material properties of powertrain components, such as coatings, surface texture, and engine lubricant viscosity. Besides improved design by the mechanical configurations, automotive engineers and lubrication scientists have applied new approaches for friction and wear control using (2) lubricant additive technologies, (3) surface engineering design. This chapter will review the surface engineering approach for friction reduction and wear control. The connection between surface engineering design and tribological performance requirements will be described to incorporate the major surface engineering processes in the emerging powertrain development trends.

In this chapter, the author will start to review the major surface engineering processes used for the automotive industry including the vapor deposition process, ion beam-assisted deposition process, and thermal sprayed process by high-velocity oxygen flame or plasma spray techniques [5–12]. Those surface coatings have been applied to vehicle powertrain vehicle components. Other surface engineering technologies such as surface texturing can affect the friction and wear behavior between the surface contacts similar to applications of surface coatings. However, reducing friction with surface texturing has the additional benefit of increasing wear resistance because of lubricant interaction with surface grooves, compared to using surface coatings alone. More details concerning lubricant and additive effects will be discussed in subsequent sections. The impact of surface coatings or surface textures on powertrain friction and wear will be reviewed. In addition, the OEMs have accelerated R&D activities to respond with novel coatings or surface textures to reduce friction and protect against wear, keeping the powertrain components working at the optimal fuel economy while retaining performance requirements for durability and reliability.

2. Surface coatings application for automotive components

In our automotive and manufacturing industries, a broad range of surface coatings used in automotive powertrain components have been applied for friction reduction and wear resistance improvement [5–12]. These surface coatings range from the application of thick deposited protection layers through chemical and physical vapor deposition or surface treatments of metal surfaces such as carburizing and nitriding in piston rings or camshafts. As shown in **Table 1**, major surface coating methods are described by their surface processing characteristics and individual preparation methods used in automotive or manufacturing applications.

2.1 Methods for preparation of surface coatings

2.1.1 Physical vapor deposition (PVD) process

The PVD process represents a thin film deposition process where a solid material as a pure metallic material or an alloy is vaporized in a vacuum environment and deposited on substrates to form a coating with the desired properties [5, 6]. As described in **Figures 1** and **2**, the metal target was heated in a vacuum and transformed into evaporation, then deposited onto a base plate with a film thickness of 5–100 nm. PVD Coating processes can be prepared by the following methods: (a) Evaporation, (b) sputtering, and (c) ion plating. Using the partially ionized metal vapor, it can react with certain gases to produce metal-based hard coatings with a specified composition on the substrate. Sputtering and cathodic arc process are the most popular methods. In the sputtering process, a metal target is bombarded with energetic gas ions to form a thin film coating. The cathodic arc method applies the vacuum arc discharges to strike the metal target and evaporate the surface material to form a PVD coating.

Coating	Preparation	Methods		
PVD	Vacuum Evaporation	Ion Plating	Hollow cathode discharge	Sputter plating
CVD	Normal Pressure CVD	Plasma CVD	Microwave plasma	Laser CVD
IBAD	Vacuum evaporate plating	Ion Implementation	Using high voltages for treatment	Using magnetic field or high voltages
Thermal Spray	High-velocity oxygen fuel spray	Plasma Spray	Electric spray	High-energy beam spray

Table 1.
A summary of the preparation methods of surface coatings and their surface processing characteristics.

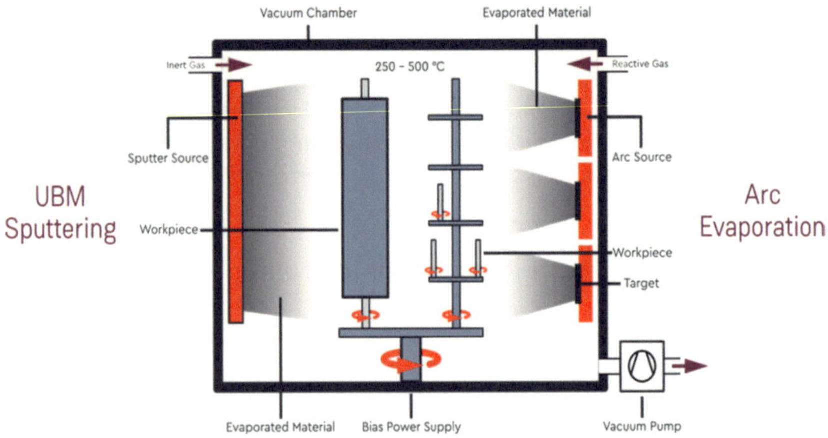

Figure 1.
PVD process.

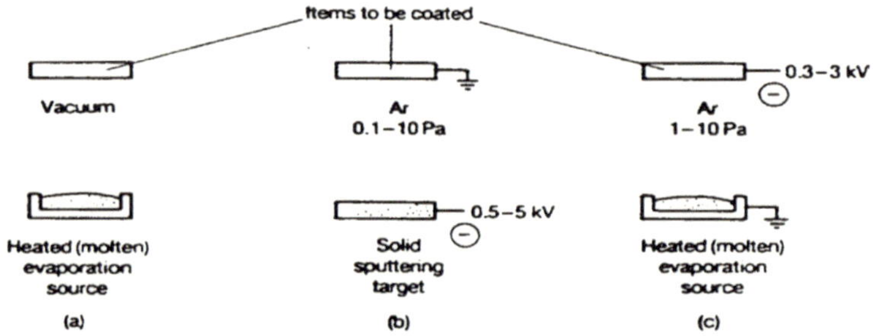

Figure 2.
PVD processes (a) evaporation, (b) sputtering, (c) ion plating.

These processes provide crucial performance attributes for the final coating properties. The nano-structured and superlattice variations of multi-layered coatings can enhance coating properties such as the desired mechanical properties in terms of friction reduction, hardness, and surface adhesion. The PVD coatings can be produced into extremely thin or very durable coating, such as hard protective films or wear resistance platings. If the PVD process transfers the coating material on the molecular level or multi-layered coatings, it can provide high-performance coatings for automotive applications to improve friction reduction and wear control.

Automotive components need to operate under severe operating conditions such as high contact loads or harsh environments with the presence of abrasive particulates or contaminants. In heavy machinery industrial applications, The PVD coatings are useful for improving tribological performance. For example, surface coatings made of nitrides, carbides, and carbonitrides of titanium (Ti), chromium (Cr), zirconium (Zr), as well as aluminum-chromium alloy (AlCr), aluminum-titanium alloy (AlTi), titanium-silicon alloy (TiSi) have been applied on a broad range of manufacturing applications Besides automotive applications, it is used in a wide variety of industry applications ranging from semiconductor devices, photovoltaic, and manufacturing tools applications.

2.1.2 Chemical vapor deposition (CVD) process

Chemical Vapor Deposition (CVD) is a method used to produce high-quality coatings by thermally induced chemical reactions [6–7]. In the CVD process, a substrate is exposed to the volatile precursors reacting with these precursors, and decompose on the substrate, creating the desired coating. Volatile by-products can be eliminated by gas flow through the reaction chamber. In general, the CVD process requires high depositing temperatures and vapor pressure. CVD Processes can be classified into two types: (A) Hot-wall thermal CVD, and (B) Plasma-assisted CVD, as shown in **Figure 3**. To ensure that the deposit was adhesive to the base plate, the saturated vapor pressure of the deposit should be maintained at an extremely low level during the deposition process.

2.1.3 Ion beam assisted deposition (IBAD) process

Ion-beam-assisted deposition, or ion-assisted deposition, is a thin-film deposition method involving both ion bombardment and physical vapor deposition (PVD) processes. The IBAD process takes place in a vacuum environment [8]. IBAD techniques combine vacuum vapor sputter with brunching ion implementation as shown in **Figure 4**. While the PVD technique evaporates the material, an ion source directs

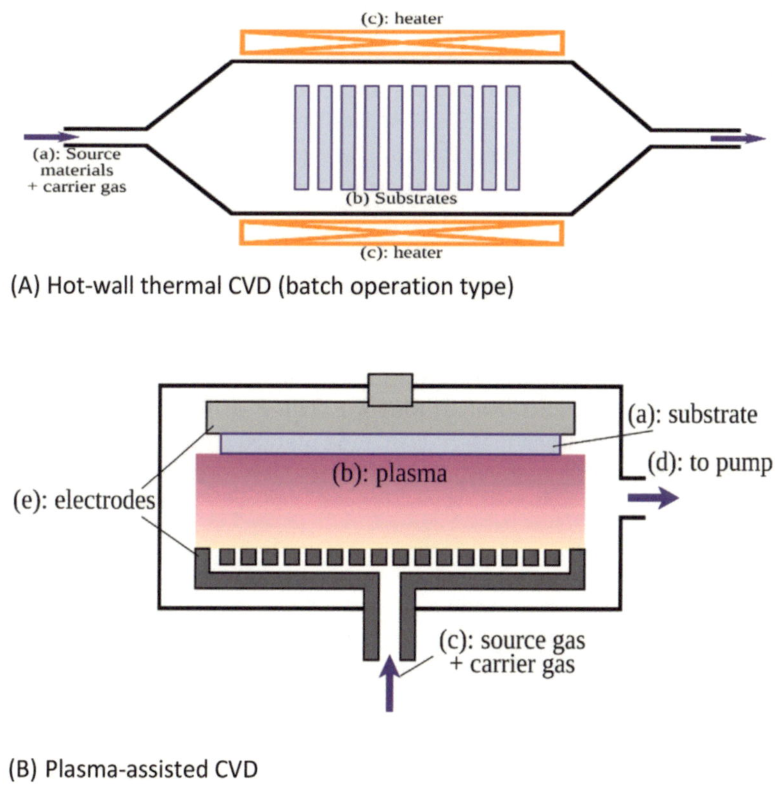

Figure 3.
CVD process: (A) Hot-wall thermal CVD (batch operation type), (B) Plasma-assisted CVD [6, 7]. (A) Hot-wall thermal CVD (batch operation type). (B) Plasma-assisted CVD.

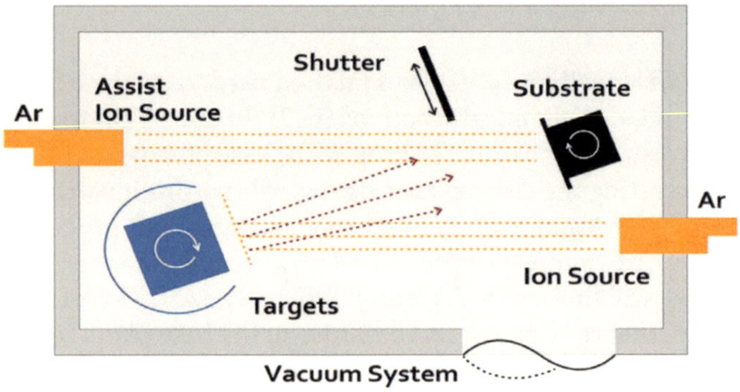

Figure 4.
IBAD (ion beam assisted process).

high-energy nitrogen and argon ions at the growing thin film. An ion-beam-assisted deposition system provides independent control of several deposition parameters including the ion energy, the substrate atoms' arrival rate, and the temperature control. The ions are then accelerated, focused, or deflected using high voltages or magnetic fields. This accelerated energy usually ranges from a few eV up to a few keV. The IBAD process can produce thin films with thickness from 0.2 um to tens of micrometers. Major applications include the deposition of ceramic or metallics materials, including gold, silver, titanium, and platinum. Ceramics include aluminum oxide, silicon dioxide, and titanium nitride coatings. Machine tool manufacturers often add a TiN coating to milling drill bits, improving these products' edge retention and corrosion resistance [8]. In addition, IBAD is the ideal coating process for industrial applications that require precision tuning of a coating's thickness and reflective index to produce a wide range of surface coatings for manufacturing or optical applications.

2.1.4 Thermal sprayed coating process

The thermal spraying coating process is a large-scale industrial coating process in which melted (or heated) materials are sprayed onto a surface [9, 10]. The "feedstock" is heated by chemical means (combustion flame) or electrical (plasma or arc) as shown in **Figure 5**. Feedstock materials include metals, alloys, ceramics, plastics, and composites. The coating materials are fed in powders or the wire forms which

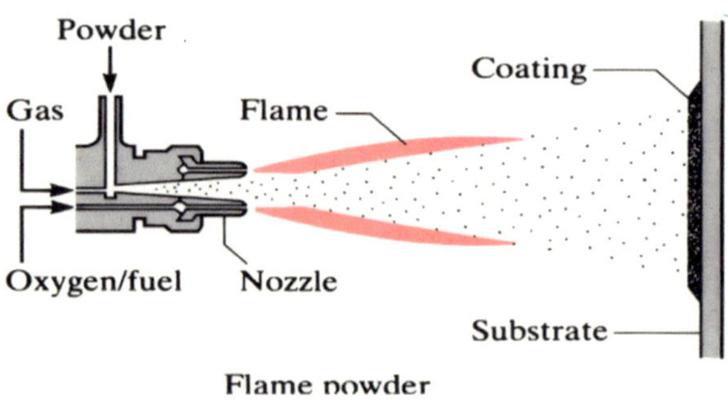

Figure 5.
Thermal sprayed coating process [9, 10].

are heated in a molten state and accelerated the micrometer-size particles into the substrate to form thermal sprayed coatings. Combustion or electrical arc discharge process is commonly applied to be the source of energy carriers for the thermal spraying process. Thermal sprayed coating quality is usually monitored by measuring its porosity, oxide content, bond strength, hardness, and surface roughness. The coating quality increases with increasing particle velocities but decreases with increasing oxide or porosity content.

The thermal sprayed coating structures consist of these lamellae 'splats', formed by flattening liquid droplets. The lamellae splats have a thickness in the micrometer range from several to hundreds of micrometers. Between these lamellae structures, there are small voids, such as pores or microcracks. As a result of this porous structure, the deposits on substrates can have specific coating properties different from the original bulk materials. Because of the rapid solidification, metastable phases can be present in those porous structures. Those porous coating structures can create the desired mechanical properties including higher strain tolerance and lower thermal or electrical conductivity for industrial applications.

2.1.5 High-velocity oxygen fuel (HVOF) and plasma spraying processes

Thermal sprayed processes can be classified into two different processes [9–11]. The first common process is High-velocity oxygen fuel spraying (HVOF) which has been applied to manufacturing industries. For example, thermal barrier coatings can be applied for exhaust heat management, wear resistance, and protection from corrosion or erosion. In addition, these coatings can change the electrical or tribological properties of the mating surface and replace worn material after long-term use. High-velocity oxygen fuel (HVOF) and plasma spraying processes have been developed for the automotive and aerospace industries [9, 11]. In the HVOF process, a mixture of fuels and oxygen gas diffuses into a combustion chamber, where the mixtures are ignited by an ignition plug and combusted into a chamber as shown in **Figure 6**. The fuels can be hydrogen, methane, propane, propylene, or natural gas. The resultant thermal sprayed particles at a pressure close to 1 MPa propagate through a converging-diverging nozzle and travel through a barrel region. At the exit of the barrel region, the jet velocity (>1000 m/s) can exceed the speed of sound. A powder feed stock is injected into the gas stream, which accelerates the powder up to 800 m/s. Due to the high-velocity gas stream, the powders are partially melted in the stream and deposited on the target substrate. The coating properties possess low porosity and high bonding strength.

The second process is the plasma spraying process [10, 11]. It has been widely used in automotive powertrain components or tool manufacturing applications.

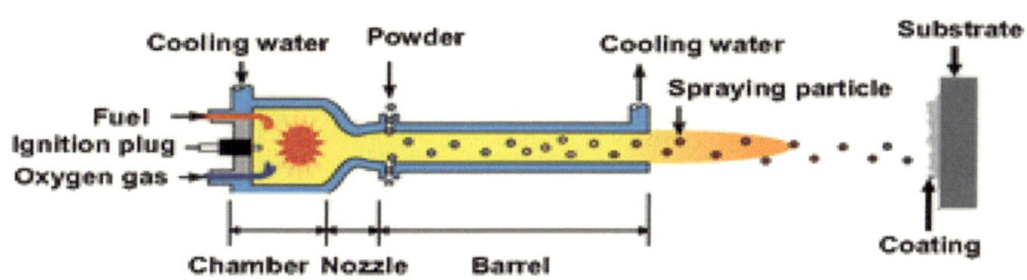

Figure 6.
High-velocity oxygen fuel spraying (HVOF) [9].

In addition, a more advanced plasma spraying process with plasma jet as shown in **Figure 7** has been used to produce wear and corrosion-resistant coatings on engineering materials, such as stainless steel, nickel-based, aluminum, and ceramic materials for industrial applications. Such coatings protect against high temperatures operating conditions. The plasma spray process has been used successfully for the desired materials such as WC–Co, chromium carbide, and alumina. The feedstock materials to be deposited are introduced into the plasma jet In the plasma spraying process. In the nozzle region, as shown in **Figure 7**, where the temperature can reach about 10,000 K, the feedstock materials are melted, and spraying particles are injected onto a substrate. The molten droplets flatten, rapidly solidify, and form deposit films on the substrate. Usually, the deposit films remain on the substrate as adherent coatings. Coating processing operating parameters influence the interaction of the particles with the plasma jet and change the coating properties. These operating parameters in the thermal sprayed process include feedstock type, plasma gas composition, substrate cooling rate, energy input, and the applied flow rate. By adjusting these operation parameters, the coating properties can be optimized according to the requirements of engineering applications.

2.1.6 Diamond-like coatings or carbon films (DLC)

A very recent class of diamond-like coatings has become particularly important for the automotive manufacturing industry. Diamond-like coatings or carbon films [12–14] are described as a technique for depositing DLC quasi-amorphous materials, which have been applied to engineering applications such as friction reduction films on aluminum or chromium materials. This technique has wide applications because of its desirable properties, in which an amorphous silicon layer about 2–4 nm thick is deposited the metal substrate, forming an interface possessed with low friction coefficient and chemical inertness. DLC's chemical bonding scheme gives it a combination of useful properties. Materials in the DLC coatings family have hardness ratings that range almost up to diamond hardness. Besides friction reduction, these coatings are

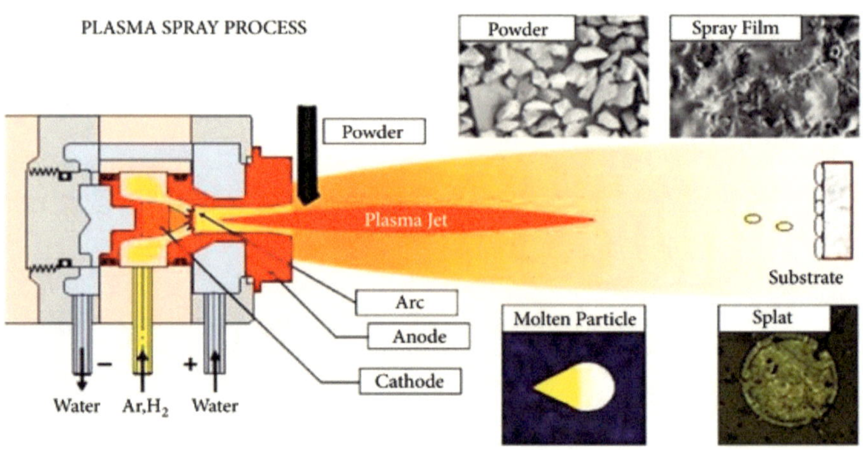

Figure 7.
Plasma spray process [10, 11].

wear-resistant materials. The DLC coatings are pretty smooth, and it can undergo surface transformations under stress to reduce friction. DLC coatings with metastable structures are stabilized by residual stresses formed during coating deposition. DLC is usually deposited on the silicon substrate which is used as a protective coat for moving parts. The films contain a transition coating between the metal and the silicon, while the DLC adheres to the silicon substrate. It has been widely applied to powertrain components for friction and wear reduction.

Recently automotive industries have applied DLC coatings in engine valvetrains and pistons for friction and wear reduction. DLC coatings can form very smooth and hard coatings with extremely high wear resistance and low friction for powertrain components. In addition, advanced powertrain components such as turbocharged engines or gasoline direct injection (GDI) engines have applied diamond-like carbon or thermal sprayed coatings for friction reduction and wear protection. CrN and TiN coatings have similar properties to DLC and have good potential for wear protection [6–8]. Those coatings have been widely applied to piston rings or valve train components because of their relatively lower cost and versatile manufacturing applications [9–11].

To make the DLC coatings adhere to the metals, Harris, Tung, and Simpco [12] have applied an amorphous SiN coating as an interlayer coating for drivetrain components because of the SiN coating's strong adherence to the metal substrate and the DLC coating adherence to the SiN. In our laboratory bench testing after lubricated sliding for 30 hours, the DLC/SiN-coated substrate showed substantial friction reduction and improved wear resistance compared with either uncoated or SiN-coated substrates without DLC coating. In addition, The DLC/SiN-coated plates were analyzed using X-ray photoelectron spectroscopy and sputter depth profiling. Based on surface profilometry analysis, the DLC/SiN coated plate showed a much smoother surface with good wear resistance. Their surface analysis results using X-ray photoelectron spectroscopy demonstrate that DLC is a promising wear protection coating for engineering applications. The commercial DLC/SiN coated parts have been applied for rolling bearing or valvetrain components to improve tribological characteristics such as high wear resistance, low friction, good embedability, better corrosion resistance, and excellent durability. In addition, Erdemir and Fenske [13, 14] have examined the properties of DLC films and their coating preparation methods. The plasma deposition process can further form exceptionally hard coatings with extremely high wear resistance and low friction.

Table 2 lists the major types of tribological coatings used for powertrain components. These tribological coatings are compared in their main applications, advantages, and disadvantages in the following **Table 3**.

In general, surface engineering can be classified into two diverse groups (a) surface modification by using either composition changes or without changes; (b) surface coatings by using either the plating & anodizing processes or vapor phase deposition processes as shown in **Figure 8**. Several methods can be applied regarding composition changes, such as thermochemical, carburizing, nitriding, or ion implantation processes. Regarding composition unchanged processes, major industrial processes such as topographical modification, transformation hardening, laser hardening, or induction hardening process have been used. **Figure 8** shows a general classification of "surface engineering", to encompass all coating materials and its preparation methods which modify and optimize solid surfaces for engineering applications [2, 8].

Coating process type	Typical coating treatment process	Typical substrate/engineering applications
Electroplating and Anodizing Process	Nickel polytetrafluoroethylene (PTFE) plating, hard chromium plating, and hard anodizing processes.	Engine pistons, piston rings, piston skirts, valvetrain, and camshafts.
Plasma or Fusion process	Plasma or fusion sprayed, and HVOF thermal sprayed coatings.	Aluminum engine bores, valve trains, or aircraft control bearings.
Vapor Phase Process	Chemical vapor deposition, physical vapor deposition, or diamond-like carbon coatings.	CVD, PVD, or DLC coatings for powertrain or drivetrain components. DLC/SiN, ion plating, and TiN coatings on the engine or transmission clutch components.
Automotive Applications	**Ceramic coatings (TiN, TiC, TiAlN, TiAlC, CrN, ZrN, etc.)**	**Diamond-Like Carbon (DLC)**
Automotive components—Engine and Valvetrain	Cylinder bores, piston rings, piston skirts, valve train components, camshafts, rolling element bearings, etc.	Fuel injectors, piston rings, turbocharger parts, and journal bearings.
Automotive components—Drivetrain	Drivetrain, transmission gears, planetary gear sets, clutches, and rear axle differential hypoid gears.	Synchronizing rings, drivetrain, and automotive bearings.

Table 2.
The major tribological coatings and applications used for powertrain components.

Coating process	Main applications	Advantages	Disadvantages
PVD	Optics, electronics, lubrication, automotive components, decorating, energy, etc.	Good surface roughness, good compatibility, and good interfacial materials.	Adhesion weakness, and bonding strength.
CVD	Electrical, automotive, anticorrosion, energy, batteries, etc.	High adhesion, high hardness, good bonding strength, good compatibility, and diffusion interface	High deposition temperature, polluted environment, and low efficiency.
IBAD	Manufacturing tools, automotive, and surface modification, etc.	A low temperature of base plate, high reliability, high repeatability, and high hardness.	Must prepare under vacuum condition, low efficiency, and high capital cost.
Thermal Spray	Manufacturing process, automotive, rapid molding, and tools, etc.	Being able to spray any metals or ceramics materials, high-velocity spray process, high deposition efficiency, and high adhesion.	Need high deposition temperature, high porosity materials (>3–15%), high capital equipment, and complicated processes.
HVOF	Automotive, manufacturing, tools, electrical and aerospace, etc.	Supersonic particles, high deposition efficiency, excellent adhesion, low porosity materials (<2%), low oxidation, and high wear resistance.	High-temperature process, high capital equipment, and complicated sprayed processes.

Table 3.
The surface coating processes are compared by their main applications, advantages, and disadvantages.

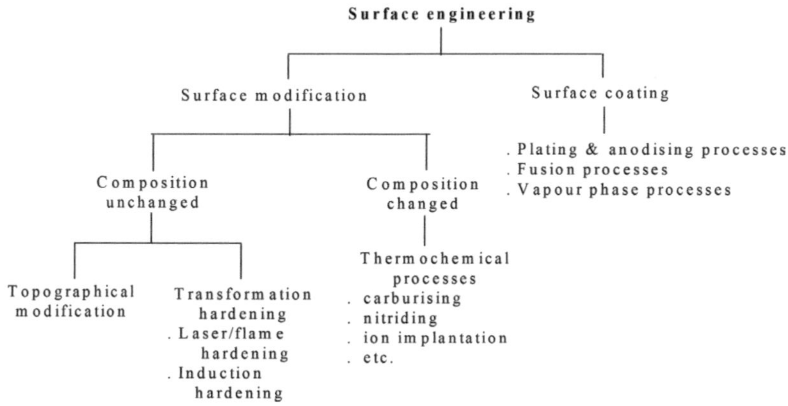

Figure 8.
Classification of surface engineering.

3. Surface coatings for powertrain applications

3.1 Advanced lightweight materials and coatings of powertrain or propulsion components

In the automotive industries, the applications of lightweight materials have been widely used to meet the increasing demand for improved energy efficiency, reduced friction, and extended wear life of engine and transmission components. These automotive new trends have stimulated industrial investigation using lightweight materials in tribological applications such as aluminum engines or metal matrix composites for brakes or heavy-duty trucks for fuel economy and durability improvement. Recently Department of Energy (DOE) made announcements regarding the current vehicle technology program (VTO) using lightweight materials applications for automotive industries [15]. DOE has found that using lightweight materials for one-quarter of the U.S. fleet could save more than 5 billion gallons of fuel annually by 2030 [15]. In addition, lightweight materials are critically important for hybrid electric, plug-in hybrid electric, and fully electric vehicles. Using lightweight materials in these vehicles can justify the weight reduction of electrification systems such as batteries and electric motors, improving energy efficiency and increasing their driving distance with extended durability. Using lightweight materials with another benefit could enable engineering applications of the compact size electrification components while promoting the all-electric or plug-in vehicles more effective for better driving performance.

DOE has made interesting forecast that lightweight engineering materials such as aluminum, magnesium, or glass fiber-reinforced polymer composites will replace heavy steel or cast iron components in the near term. Those lightweight materials in powertrain components can decrease component weight by 30-60%. Industrial engineers are investigating the advanced manufacturing processes to further improve energy efficiency by 40–50% using superplastic metal-forming (SPF) processes for joining, metal forming, and recycling these materials [15].

Surface coatings for powertrain components enable different lightweight materials or alloys to be used in powertrain parts, which can provide friction reduction and wear resistance benefits [16]. Friction and wear control of powertrain components can be approached in three major areas: surface engineering modifications, lubricant additives interaction, and mechanical component design between the rubbing surfaces and lubricant films. In recent years, significant advances have been made in lightweight material development and surface engineering applied to powertrain components. Recently lightweight non-ferrous alloys have been incorporated into advanced manufacturing processes using thermally sprayed aluminum liners, DLC-coated piston rings, or rolling element bearings for powertrain and transmission components. These advanced coatings significantly contribute to friction and wear control in advanced powertrain or hybrid vehicles.

3.2 Impact of thermal spray coatings on non-ferrous engine blocks for reducing friction and wear

The automotive industry is dealing with tough overseas competition, government regulations, and rapid technological changes. Higher-performance engineering materials and longer durability of powertrain systems will become increasingly important in the face of harsh operating environments and stringent federal legislation calling for better fuel economy and lower emissions. Presently the engine lubricant formulations are specifically designed for cast iron blocks, not for non-ferrous engine blocks such as aluminum engine blocks or magnesium materials. The current engine lubricant formulations contain antiwear additives such as zinc dialkyl phosphates which are incompatible with non-ferrous engine blocks. The current manufacturing process of aluminum engines requires cast iron sleeves to interact with lubricant additives. These manufacturing processes for cast iron sleeves are very expensive for machining. To reduce manufacturing costs by using cast iron sleeves, automotive industries have applied thermal sprayed coatings on the top of non-ferrous engine blocks to improve the wear resistance of aluminum engine blocks. Hartfield and Tung [17] have developed thermal sprayed iron oxide coating on the top of aluminum engine bores to provide a smooth surface with good lubrication characteristics [17]. Those thermal sprayed iron oxide coatings on aluminum bores have replaced the conventional cast iron bores or liners because the thermal sprayed coatings have reduced engine weight, better heat transfer properties, improved wear resistance, and lower machining costs than the cast iron blocks or liner sleeves [17].

In this thermal sprayed coating process, the high-velocity oxygen fuel (HVOF) method was applied to the engine cylinder block with a robust manufacturing process as shown in **Figure 9** [17]. In the first treatment of this manufacturing process, a 30–40 KSI water jet was sprayed into the base aluminum to roughen the surface and promote the coating adherence to the base aluminum. During the second step under an environment of methane and excess oxygen, a low-carbon, 1010 steel wire was heated and sprayed onto the surface. This thermal sprayed process can generate a uniform coating with a volume composition of 50% Fe and 50% FeO. After the spraying process, another water jet cleaning process was applied to clean all loose particles on the honed surface. Using this effective water jet cleaning process can remove torn and folded metals or loose particles. Due to the flattening of the molten metal, thermal spray microstructures consist of multiple layered structures called "splats". The wear mechanism is usually referred to as splat delamination which needs to be prevented for wear propagation by using effective lubricant additives. Lubricant additives can help protect this wear propagation and extend the wear life of thermal sprayed coatings.

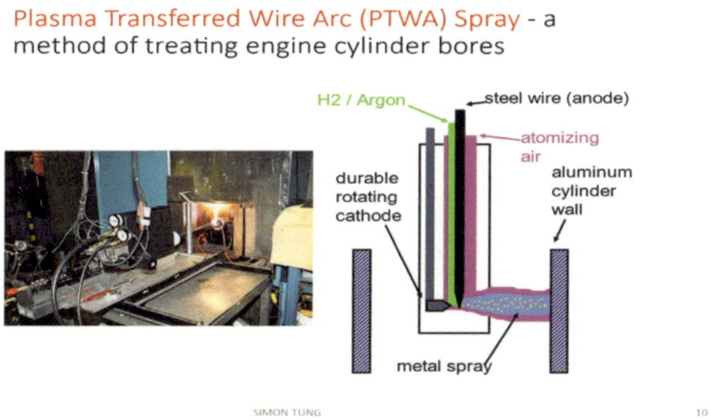

Figure 9.
Plasma transferred wire arc (PTWA) spray—a robust method of treating engine cylinder bores [17].

Tung and Wong [16] have applied the plasma spray Fe_2TiO_5 coatings on aluminum engines for friction and wear reduction using the commercial available engine oil containing friction-reducing additives MoDTC (molybdenum dithiocarbamate). These friction-reducing additives interacting with the plasma spray Fe_2TiO_5 coating will further reduce both engine cylinder friction and piston ring wear. They have modified the Cameron-Plint High-Frequency Friction and Wear Test Machine to test the aluminum engine bores coated with the plasma spray Fe_2TiO_5 coatings after lubricating in engine oil containing MoDTC. In addition, the plasma spray Fe_2TiO_5 coatings on aluminum engine bores have been tested against a molybdenum-sprayed piston ring. The plasma spray Fe_2TiO_5 coating also demonstrated a low friction coefficient and better wear resistance than the uncoated aluminum cylinder bore. The plasma spray Fe_2TiO_5 coating against the aluminum engine bores showed a low friction coefficient $\mu_f \approx 0.11$ @ 410°C. The wear volume was lower than 0.02 cm^3 after 50 hours of lubricated sliding. The uncoated aluminum cylinder bore showed much higher bore wear at approximately 1.5 cm^3 after 50 hours of lubricated sliding. All thermal sprayed coatings on aluminum bores lubricated with engine oils containing friction modifier MoDTC have significantly improved wear resistance and friction reduction. Both lubricant additives and thermal sprayed coatings have synergized effect on friction and wear reduction.

3.3 Piston skirt/cylinder bore scuffing

Piston skirt/cylinder bore scuffing is one of the major wear modes contributing to the failure of automotive engines. Scuffing phenomena can be observed by dramatic friction increase and rapid temperature rise that causes lubricant degradation. The scuffed parts in the piston skirt/cylinder bore interface may cause vibration and noise, or even seizure of the interface. Scuffing wear is heavily involved with thermal, physical, and chemical interactions among the contacting bodies, the environment, and the lubricant additives at a sliding interface. Wear mechanisms are mainly due to lubricant breakdown and the loss of successive protective films. Surface interactions in mixed or boundary lubrication regions initiate the material plastic flows in the surface contact region. Such interactions directly impact the change of material microstructures and generate surface cracks before the final adhesive phase. The adhesive wear process substantially influences the scuffing progression. After the

Figure 10.
Modified bench testing for piston skirt coatings to simulate the relative motion between a piston and a liner of typical engine operating conditions.

final stage of scuffing process, metallic adhesion of the protective films occurs over large surface contact areas [18, 19].

To understand the scuffing mechanisms of the piston skirt/liner contact of automotive engines, researchers at General Motors Powertrain [18, 19] have explored the application of surface coatings on piston /liner contact for scuffing prevention. They used a variety of coating material pairs to investigate the piston skirt scuffing wear under the reciprocating motion environment similar to that in a real engine. In their publication [18, 19], a piston scuffing apparatus was developed to simulate the relative motion between a piston and a liner of typical engine operating conditions as shown in **Figure 10**. The scuffing performance of a coated piston with a variety of surface coatings was investigated using a custom-designed tribology apparatus. In addition, the impact of surface texture effect on the cylinder liner–piston interface was studied and tribological characteristics were determined in the R&D laboratory [19, 20].

Wang and Tung have conducted both the bench scuffing test and engine dynamometer tests [18, 19]. The bench scuffing tests were conducted at both hot and cold scuffing conditions. In the bench simulation tests, a variety of surface coatings including the iron plating, Ni-W plating, electroless Ni plating, DLC, hard anodizing, as well as nickel composite coatings (Ni-P-SiC, Ni-P-BN, and Ni-P-Si_3N_4) have been tested using the 2.5-hour cold scuffing and 9-hour hot scuffing testing conditions.

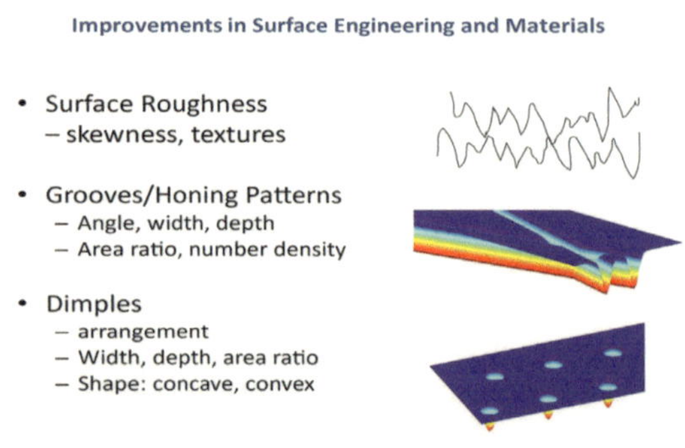

Figure 11.
Improvements in surface engineering and materials by surface roughness, grooves, and dimples.

Testing results of the tin-coated pistons against the 390, 413, or 356 aluminum engine bore indicated that the bore polishing with severe bore wear was observed after a cold or hot scuffing test. This might be due to the poor lubrication condition between the piston and engine bore at either cold or hot temperatures. However, the Ni-P-BN coating, iron plating, and modified hard anodizing coatings on piston skirts did not show piston scuffing and cylinder bore wear. All tested pistons and aluminum bores passed both the cold and hot scuffing bench tests. The bench scuffing simulation testing results as shown in **Table 4** can be summarized into the following five groups (I, II, III, IV, and V):

I. This group of coatings could only provide a "break-in" function not wear protection, such as a tin-coated piston.

II. The second group coatings such as the TiN coating, hard anodizing, and Ni-W plating pistons showed abrasive wear after the cold and hot scuffing dynamometer tests, which might result from abrasive aluminum oxide particles coming off the anodized piston skirts during sliding. The second group coatings did not pass through scuffing tests at either high loads or high temperatures operating conditions;

III. The third group coatings including the DLC, Ni-P-SiC, Ni-P-Si_3N_4 coatings, and the electroless nickel-coated pistons indicate mild wear on the aluminum cylinder bores. They have provided much better protection and showed more durability after lubrication starvation occurred at extreme hot or cold temperatures.

Scuffing test results	**Bore materials**			
Piston Coatings	390 Al	413 Al	356 Al	Cast Iron
Sn-coated skirt	I	I	I	I
Ni-P-BN	IV	IV	IV	IV
Ni-P-SiC	III	III	III	III
Ni-P-Si_3N_4	III	III	III	III
Fe-plated skirt	V	V	V	V
D10 (DLC Coating)	IV	III	III	III
Modified Hard Anodizing	IV	III	III	IV
HMB-2	II	II	II	II
HMB-31G	II	II	II	II
HMB-4A	II	II	II	II
Hard Anodizing + HMB-2	II	II	II	II
Hard Anodizing + HMB-4A	II	II	II	II
Hard Anodizing + HMB-31G	II	II	II	II
Hard Anodizing + D-10	II	II	II	II

Note: Ranking order I, II, III, IV, and V, have been described on the last page by their scuffing test performance.

Table 4.
Tested piston and bore material combinations conducted in bench scuffing tests [19, 20].

IV. The fourth group coatings including Ni-P-BN, D10, and modified hard anodizing coatings could survive at reasonably high loads or hot temperatures against the 390, 413, or 356 Al bore samples. This coating group has better scuffing resistance compared with the last groups tested under either extremely hot or cold environments.

V. The fifth group coatings including the Fe-plated skirt coating, and modified iron electroplating coating, had excellent wear resistance performance in terms of its scuffing resistance against the 390, 413, 0r 356 Al bores.

Engine dynamometer tests have been conducted for these tested piston skirt coatings against the 390, 413, or 356 aluminum engine bores. Engine dynamometer testing results showed the same ranking as the simulation bench test results in terms of scuffing and wear performance. In addition, the bench wear simulation test results were correlated well with the engine dynamometer test results. Therefore, it has proved that the bench wear simulation test can be used as a rapid, low-cost, and reproducible method for screening the tribological characteristics of all piston skirt coatings.

In addition, Wang, Tung, and Vernier [20, 21] have discovered a composite polymer coating (CPC) over a hard anodized surface for reducing both wear and scuffing. Experimental observations for piston skirt coatings indicated that the composite polymer coating over a hard anodized surface with about 15 vol.% graphite particles. The modified hard anodizing coating with CPC had the highest scuffing resistance compared with the other piston skirt coatings including a nickel-phosphorus plated coating with 4 vol.% boron nitride particles (NCC), or using a boron nitride coating alone. The piston's hard anodizing surface coating improved surface adhesion with the composite polymer coating and provided with additional solid support to the composite polymer coatings in terms of scuffing resistance. Another discovery of these CPC piston coatings showed a synergistic effect of wear protection if the engine lubricant additives contain friction-reducing molybdenum disulfide additive or graphite particles dispersed in CPC coatings. Wang and Tung invented a unique composite polymer coating over a hard anodized surface specially designed for aluminum engine components for wear and scuffing resistance. This invention applied a composite polymer coating atop the hard anodized layer directly on the aluminum engine bores. This CPC coating has significantly improved wear and scuffing resistance compared with the other commercial coatings in the global market.

4. Powertrain friction and wear reduction by applications of surface textures

Besides the design of surface coatings for different powertrain configurations, the use of surface texturing has been applied for friction and wear control in the literature [16, 22]. While surface coatings can protect the surface from abrasive or adhesive wear by adjusting the properties of coating materials, surface texturing can influence friction and wear characteristics through surface grooves or honing patterns. It will be the major focus of our discussion in this section.

4.1 Surface textures

Surface texturing has become a vital surface engineering technique for controlling surface properties in specific applications such as anti-wear, friction-reducing, self-healing, or anti-biofouling properties Recently several research papers [22–25] claimed that the design of surface texturing on engineering materials can improve the tribological performance under different operating conditions such as adhesive wear, cavitation wear, and abrasive wear environment. Automotive industries have widely adopted this surface texturing technique to enhance tribological characteristics such as wear and friction. Previous studies have explored the potential of surface grooves or honing patterns to provide film lubrication, while the role of surface texturing has demonstrated more interesting surface interaction between metallic surfaces and lubricants in surface grooves or honing patterns. The surface interaction by lubricant additives or surface texturing can create a balance between hydrodynamic and boundary lubrication. It also reduces the amount of asperity contact that takes place during the lubricated sliding conditions. When surface contact does not occur, an increase in oil film thickness can reduce oil shearing and hydrodynamic friction. As shown in **Figure 11**, three methods such as surface roughness, grooves/honing patterns, or dimples can help improve tribological characteristics of surface engineering and materials [22–25].

4.2 Impact of cylinder bore surface finishes and honing

Surface texturing treatment is an emerging technology that potentially impacts lubrication, friction, and wear for powertrain components [22–25]. Details of surface texture designs have been described including surface roughness, grooves/honing patterns, and surface dimples as shown in **Figure 11**. Among these surface texturing treatments, engine cylinder bore honing patterns were proved to be the most effective process in engine surface modification. Their studies also verified that if the cross-hatch honing patterns aligned in a direction more perpendicular to the sliding motion, these honing patterns can provide much better lubrication benefits compared with a direction aligned with more tilted angles to the sliding motion. The viscosity-temperature dependence was kept the same as the overall viscosity was changed. Oil viscosity effect can help the surface texturing treatment reduce friction further and wear in all cases. The synergistic effect of both lubricant viscosity and surface texturing can enhance friction reduction as evidenced by the recent investigation.

Surface texturing enables surface modification by surface roughness changes, micro grooving, micro dimples, and microchannels. This method can be processed by laser treatment or micromachining process. Besides surface treatment, another useful technique is applied by surface honing patterns in automotive engine block machining processes, In automotive engine manufacturing processes, a major step for honing the engine cylinder bores is to produce cylinder blocks with the specific honing patterns that are close to the engine cylinder "break-in" condition. Lin and Tung [22] have discovered a unique plateau honing process to create a smooth bore surface finish pattern, which is accomplished by using the honing sticks with very fine abrasives to remove the surface peaks from the surface finish process. This process can be applied to smooth surface peaks and eliminate the torn metals left by the rough honing machining process, leaving a plateau-like surface finish. This plateau honing process generates a crosshatch pattern for retention of the lubricant in surface grooves. In addition, this finishing process provides a good bearing area for the piston and piston ring to slide over and produces a boundary lubrication or a mixed film region. It will be beneficial for improving the lubricity and reduce friction and wear. The Plateau honing process can reduce oil consumption on the cylinder walls and enable engines with less trapped hydrocarbons and lower emissions. Additional benefits of smooth plateau honing finishes can substantially reduce ring wear for better piston ring sealing and longer life. Plateau honing finishes also reduce the blow-by effect and eliminate oil contamination from the metal debris using this plateau honing process on engine cylinder blocks [23–25].

There is a continuous development trend to apply surface texture for improving tribological properties. Most of them are stimulated by the high potential that surface texturing can generate micro-reservoirs for improving lubrication storage or small-scale traps to catch wear debris. Takata, Li, and Wong [26] examined the influence of surface textures for controlling friction and wear of engine components. The surface texturing effect has been investigated as an effective treatment for promoting the tribological properties of powertrain components by automotive engineers [27–29]. Kovalchenko et al. [30] studied the lubrication regime simulating the Stribeck friction curves for various surface textures using a pin-on-disk bench testing apparatus under lubricated sliding. Their testing results did demonstrate the merits of surface textures for friction and wear control. These surface texturing treatments have proved very promising techniques for engineering applications.

Another method using surface dimples can promote effective hydrodynamic lubrication, extending the surface contact regime for more severe lubricated conditions at lower sliding speeds and higher loads. Sadeghi et al. [31] have demonstrated that adding surface dimples in the end-stroke region of a reciprocating engine can reduce surface contact in this boundary lubrication area. It has proved that the surface texturing effect can significantly reduce asperity contact and extend surface lubricity. Ronen, Etsion, and Klingman [32, 33] have conducted both analytical and experimental studies to validate the effects of surface dimples on sliding friction. Their research studies have shown that adding surface dimples could decrease sliding friction in reciprocating engines due to the hydrodynamic lubrication effect on surface texturing. Siripuram and Stephens [34] investigated the influence of different dimple shapes or orientations on engine friction. They explored the impact of different shapes or orientations such as square, diamond, circular, hexagonal, and triangular cross-sections. They have found that the effect of friction reduction was by dependent of different shapes of dimples in the contact zone. Hsu [35] also investigated the orientation effects of surface dimples. His studies have validated that dimple shapes could have a similar influence on lubrication and concluded that certain dimple shapes with an orientation more perpendicular to the sliding direction could delay the onset of asperity contact. Their research studies have verified that surface texture design has a pronounced effect on friction reduction and load support for automotive components.

Surface texture orientation affects friction and oil film thickness as evidenced by academic researchers [36, 37]. Michail and Barber [36] examined the orientation effect of surface textures. They observed that if surface textures are more perpendicular to the sliding direction would increase oil film thickness in the contact region due to this specific surface texture orientation. In addition, Jocsak and Wong [37] also observed that engine cylinder honing grooves would reduce friction for the lower cross-hatch angles such as 0-degree or 30-degree cross-hatch angles (grooves more perpendicular to the sliding direction) compared with higher cross-hatch angle such as 90-degree. The Y-axis represents the normalized ring pack FMEP (Friction Mean Effective Pressure). Reducing honing cross-hatch angle reduced predicted friction as shown in **Figure 12**. In addition, Jocsak, and Wong demonstrated similar results of the cylinder liner surface texturing effect on piston ring/cylinder liner friction [37]. As shown in **Figures 12** and **13**, testing results of friction mean effective pressure using three examples of surface textures (from a 30-degree cross-hatch angle to a

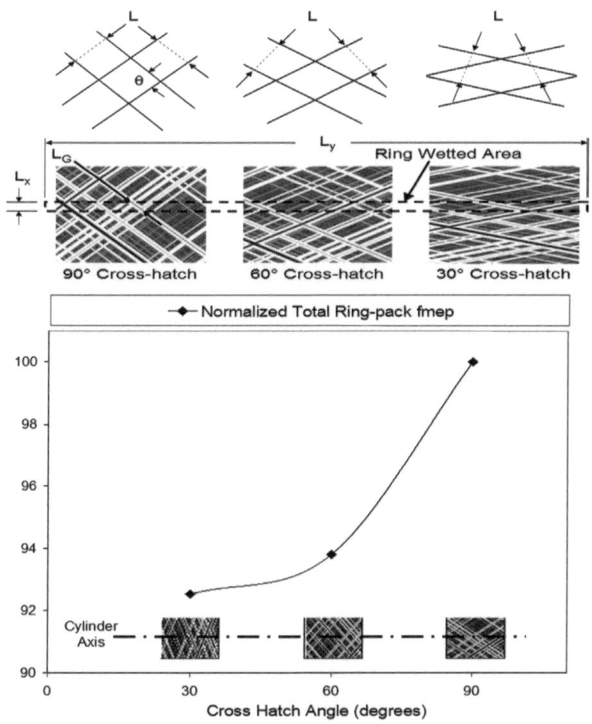

Figure 12.
Normalized FMEP (Y-axis) measurement and comparison of different cross-hatch angles (X-axis) [2, 22].

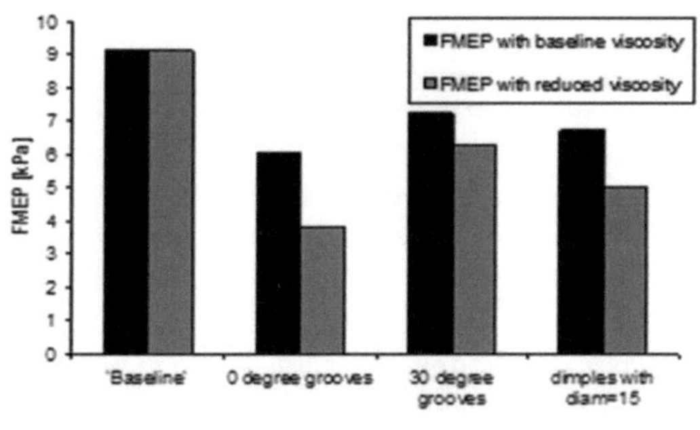

Figure 13.
FMEP (friction mean effective pressure) reduction due to the combined lubricant and surface texturing effects [2, 37].

90-degree cross-hatch angle) have indicated a positive trend of friction reduction, compared to the baseline (without surface texturing). Regarding the baseline case in their study, the cylinder liner is untextured without a cross-hatch pattern. These friction mean effective pressure testing results also demonstrated that adding a surface texturing effect can influence friction reduction. Lower viscosity lubricants combined with surface texturing patterns have indicated that FMEP would further reduced due to the combined reduced lubricant viscosity with surface texturing effects compared with a higher baseline without surface texturing.

Wong and Tung [38, 39] have applied surface textures or surface honing treatments to demonstrate the substantial benefits in the engine or valvetrain components.

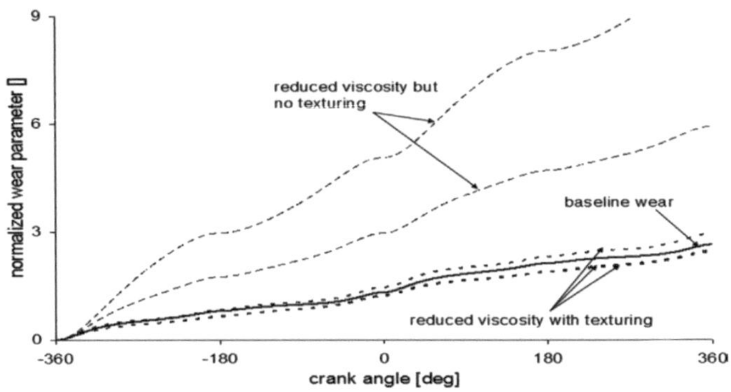

Figure 14.
The combined lubricant viscosity and surface textures affect engine cylinder wear [37, 38].

Regarding powertrain components such as valve train systems and cam lobes, wear is more sensitive to the mechanical load between contacting surfaces due to the higher metal-to-metal contacts in the valve-train system. These surface textures in the valve train system or cam lobes can affect the lubricant flow between the sliding surfaces with net results similar to the change in lubricant viscosity. Surface textures in powertrain components can generate substantial benefits in the engine cylinders, valve trains, and rolling element bearings.

Besides the surface texturing effect on engine friction, the engine testing results also concluded the significant wear resistance benefit compared to reducing friction by low viscosity lubricant alone as shown in **Figure 14**. By combination of reduced lubricant viscosity with surface texturing treatment can reduce wear substantially compared with reduced viscosity alone without surface texturing. Recently the author and his industrial colleagues [40] have developed unique engine sequence tests to evaluate the tribological characteristics of a variety of surface coatings combined with surface textures to comply with the current GF-6 engine oil specifications [40]. They have found in their engine testing results that the plateau-smooth finish pattern enables better lubrication because of the production pistons or rings sliding over production engine bores for better retention of the engine lubricants in surface honing grooves. These engine testing results have demonstrated that using surface honing patterns or textures on engine cylinder bores can produce both friction-reducing and wear-resistant benefits.

5. Conclusion

This chapter has emphasized the effects of surface coatings and surface textures on powertrain friction and wear. The author has reviewed recent publications using surface engineering methods for automotive powertrain components. In addition, their surface characteristics and individual preparation methods have been described.

Recently lightweight non-ferrous alloys have been applied to the emerging powertrain technologies for improving energy efficiency and fuel economy. A broad range of surface coating materials have been applied to improve energy efficiency and wear resistance in powertrain or propulsion systems. The significant benefit of using these surface engineering technologies on friction reduction and wear resistance improvement has been reviewed and discussed. In addition to the surface coating design for powertrain components, the applications of surface textures can further reduce friction and wear. Surface coatings can protect the sliding components from abrasive or adhesive wear through the hardness of the material or lubricating films. While surface texture modification or engine honing patterns can control both friction and wear of the contact region.

In summary, this chapter has reviewed the influence of emerging surface engineering technologies on tribological performance for advanced powertrain components. This chapter also addresses the surface interactions between surface coatings or textures and lubricated components. The increasing importance of surface coatings or textures combined with different powertrain component designs has been fully emphasized in this chapter. In addition, the OEMs have accelerated R&D activities to respond with novel coatings or surface textures to reduce friction and protect against wear, keeping the powertrain components working at the optimal fuel economy while retaining performance requirements for durability and reliability.

Author details

Simon C. Tung
Independent Researcher, Chicago, United States of America

*Address all correspondence to: simontung168@gmail.com

References

[1] Davis JR. Friction and wear of internal combustion engine parts. In: Blau P, editor. ASM Handbook, Volume 18: Friction, Lubrication, and Wear Technology. Materials Park, Ohio, USA: ASM International; 1992. pp. 553-562. DOI: 10.1361/asmhba0002305

[2] Wong VW, Tung S. Overview of automotive engine friction and reduction trends–effects of surface, material, and lubricant-additive technologies. Friction. 2016;**4**(1):1-28. DOI: 10.1007/s40544-016-0107-9

[3] Kapoor A, Tung S. "Modern tribology handbook", volume two, "materials, coatings, and industrial applications". In: Automotive Tribology. Boca Raton, PA, USA: CRC Press; 2003. pp. 1187-1229

[4] Holmberg K, Andersson P, Erdemir A. Global energy consumption due to friction in passenger cars. In: Tribology International. Vol. 47. S.A: Elsevier Publications; 2012. pp. 221-234

[5] Selvakumar N, Barshilia HC. Review of physical vapor deposited (PVD) spectrally selective coatings for mid- and high-temperature solar thermal applications. Solar Energy Materials and Solar Cells. 2012;**98**:1-23. DOI: 10.1016/j.solmat.2011.10.028

[6] Smith D. Thin-Film Deposition: Principles and Practice. New York, USA: MacGraw-Hill; 1995. ISBN 978-0-07-058502-7

[7] Dobkin, Zuraw. Principles of Chemical Vapor Deposition. ISBN 978-1-4020-1248-8. 2003

[8] Madhavi G, Naik K, Raghavendra CR. A review on recent approaches in the field of surface coating. Materials Today, Coating Material Proceedings. 2022;**52**:403-406

[9] Kuroda S, Kawakita J, Watanabe M, Katanoda H. Warm spraying—A novel coating process based on the high-velocity impact of solid particles. Science and Technology of Advanced Materials. 2008;**9**(3):033002. DOI: 10.1088/1468-6996/9/3/033002. PMC 5099653, and PMID 27877996,

[10] Paulussen S, Rego R, Goossens O, Vangeneugden D, Rose K. Plasma polymerization of hybrid organic-inorganic monomers in an atmospheric pressure dielectric barrier discharge. Surface and Coatings Technology. 2005;**200**(1-4):672-675. DOI: 10.1016/j.surfcoat.2005.02.134

[11] HVAF and plasma spray process | thermal spray coatings | machine part enhancement. HTS Coatings. South Plainfield, NJ, USA: A&A Coating Inc.; 2006

[12] Harris SJ, Tung S, Simko SJ, Militello MC, General Motors Research and Development Center. A diamond-like carbon film for wear protection of steel. Surface and Coatings Technology. 1993;**62**(1993):550-557

[13] Erdemir A, Fenske GR, Krauss AR, Green DM, McCauley T, Csencsits RT. Tribological properties of nanocrystalline diamond films. Surface and Coatings Technology. 1999;**121**:565-572

[14] Erdemir A, Fenske GR. Tribological performance of diamond and diamond-like carbon films at elevated temperatures. Tribology Transactions. 1996;**39**:787-794

[15] DOE Vehicle Technologies Office (VTO)'s Lightweight and Propulsion Materials Program Announcement. 2023. Available from: https://www.energy.gov/eere/vehicles/lightweight-and-propulsion-materials.

[16] Tung S, Wong V. Surface engineering development trends and impact of surface coatings and textures on automotive powertrain friction and wear control. ASTM Journal of ASTM Materials Performance and Characterization. 13 Oct 2021. DOI: 10.1520/MPC20200091

[17] Hartfield-Wunsch SE, Tung S. The effect of microstructure on the wear behavior of thermal spray coatings. In: Proceedings of the 7th National Thermal Spray Conference. 1994

[18] Tung S, Wang YC. Scuffing and wear behavior of aluminum piston skirt coatings against aluminum cylinder bore. In: Proceedings of the International Conference on Wear of Materials. 2000

[19] Wang Y, Cheng HS, Simon T, Jane Wang Q. An experimental investigation of piston skirt scuffing: A piston scuffing apparatus, experiments, and scuffing mechanism analyses. Wear Journal. 2001;14: 190-195. 10.1016/S0043-1648(03)00538-6. Advanced Functional Materials 2001;**11**(5):348-354; Wear 2000;245

[20] Wang Y, Vernier P, Tung S. Scuffing and wear behavior of aluminum piston skirt coatings against the aluminum cylinder bore. In: Proceedings of the 12th International Conference on Wear of Materials. 1999

[21] Wang Y, Vernia P, Tung Simon C-Y. US Patent 5884600, US Patent title Aluminum Bore Engine Having Wear and Scuffing Aluminum Piston. The U.S. Patent number is 5,884,600. 1999

[22] Lin YT. Honing with Abrasive Brushes. Chicago, IL, USA: SME Special Publication; 1998. MF92-132

[23] Schneider E, Lechman D, Shin S. The Effect of Bore Surface Finish and Deck-Plate Honing on Oil Consumption of the 7.4-L Mark V Engine. Philadelphia, PA, USA: SAE Transactions; 1996. SAE #962344

[24] Hill SH, Kantola TC, Brown JR, Hamelink JC. An Experimental Study of the Effect of Cylinder Bore Finish on Engine Oil Consumption. Philadelphia, PA, USA: SAE Transactions; 1995. #950938

[25] Ammosova L et al. Precise fabrication of microtextured stainless steel surfaces using metal injection moulding. Precision Engineering. Jun 2020;**6**

[26] Takata R, Li Y, Wong V. Effect of liner surface texturing on ring/liner friction in large-bore engines. In: Proceedings of ICEF06, ASME Internal Combustion Engine Div. 2006 Fall Technical Conference, November 5-8, 2006, Sacramento, California, USA, Paper ICEF2006-1525. 2006

[27] Hamilton DB, Walowit JA, Allen CM. A theory of lubrication by microasperities. ASME Journal of Basic Engineering. 1966;**88**:177-185

[28] Anno JN, Walowit JA, Allen CM. Microasperity lubrication. ASME Journal of Lubrication Technology. 1968;**90**:351-355

[29] Anno JN, Walowit JA, Allen CM. Load support and leakage from microasperity-lubricated face seals. ASME Journal of Lubrication Technology. 1969;**91**:726-731

[30] Kovalchenko A, Ajayi O, Erdemir A, Fenske G. The effect of laser surface

texturing on transitions in lubrication regimes during unidirectional sliding contact. Tribology International. 2005;**38**:219-225

[31] Sadeghi F et al. Advanced Natural Gas Reciprocating Engine: Parasitic Loss Control through Surface Modification. Cambridge, MA: ARES/ARICE Low Engine Friction Alliance Workshop; 2005

[32] Ronen A, Etsion I, Kligerman Y. Friction-reducing surface-texturing in reciprocating automotive components. Tribology Transactions. 2001;**44**:359-366

[33] Ryk G, Kligerman Y, Etsion I. Experimental investigation of laser surface texturing for reciprocating automotive components. Tribology Transactions. 2002;**45**:444-449

[34] Siripuram R, Stephens L. Effect of deterministic asperity geometry on hydrodynamic lubrication. ASME Journal of Tribology. 2004;**126**:527-534

[35] Hsu S. Integrated Surface Modification Technology Development. Oakridge, Tennessee, USA: Oak Ridge National Laboratory Seminar; 2005

[36] Michail SK, Barber GC. The effects of roughness on piston ring lubrication part I: Model development. Tribology Transactions. 1995;**38**:19-26

[37] Jocsak J, Wong V, et al. Analyzing the effects of three-dimensional cylinder liner surface texture. In: ICE2005-1333. Ottawa, Canada: ICEF; 2005

[38] Wong VW, Tung S. Friction, lubrication, and wear of internal combustion engine. In: Totten G, editor. Handbook of Tribology, Volume 18, Friction, Lubrication, and Wear Technology. new ed. West Conshohoken, PA, USA: ASTM International Publisher; 2017. pp. 899-915

[39] Tung S, Wong VW. Overview of automotive tribology and future development trends-impact of surface engineering and materials for friction and Wear reduction. In: 5th Asian Conference on Heat Treatment and Surface Engineering, Keynote Presentation, Hangzhou, China, November 12-14, 2016

[40] Girard L, Tung S, Woydt M, Dennis B. Automotive engine lubricants and testing. In: Fuels and Lubricants Handbook. West Conshohoken, PA, USA: ASTM International Publisher; 2019. pp. 753-863

Chapter 4

Surface Engineering Solutions for Corrosion Protection in CCUS Tubular Applications

Lei Zhao, Jiaxiang Ren, Timothy Ryan Dunne and Peng Cheng

Abstract

Deploying Carbon Capture, Utilization, and Storage (CCUS) is crucial for achieving the International Energy Agency's Net Zero Emissions by 2050 objectives. However, global implementation faces delays due to high capital costs, particularly from the corrosive environment induced by water and industrial impurities in CO_2 waste streams, which necessitates the use of expensive corrosion-resistant alloys. Downhole tubulars represent a major part of the project cost, so small changes in material selection can greatly affect the overall cost of the injection project. Surface engineering technologies such as coatings, platings, and liners combined with cost-efficient traditional steel, have shown promise in mitigating these costs. These technologies provide a technically feasible and cost-effective alternative to expensive alloys, reducing overall project costs and promoting the broader adoption of CCUS. This chapter provides a comprehensive review of these technologies in downhole tubular applications, covering their history, development routes, failure analysis, current issues, field experiences, and selection criteria, concluding with an outlook for future advancements and potential improvements in the field.

Keywords: CCUS, corrosion protection, coating, plating, liner

1. Introduction

Global climate change, propelled by the surge in greenhouse gas emissions, poses an imminent threat to the Earth's climate systems, manifesting in an array of adverse effects such as heightened and more frequent heatwaves, extreme weather events, rising sea levels, and disturbances to ecosystems. A prominent driver of this warming trend is the combustion of fossil fuels for energy, releasing substantial amounts of carbon dioxide (CO_2) byproduct into the atmosphere. Despite the rapid advancement of alternative energy sources such as wind, solar, and hydrogen, a considerable portion of the world's escalating energy demand is projected to continue being met by fossil fuels in the coming decades, which requires urgent deployment of technologies like Carbon Capture, Utilization, and Storage (CCUS) to mitigate emissions from large-scale fossil fuel utilization. Wood Mackenzie's forecast in the 1.5°C

Accelerated Energy Transition (AET) Scenario indicates a critical need to reduce emissions by 15–20% [1], necessitating a CCUS capacity of 5 billion tons per year. Despite substantial governmental support, notably through initiatives like 45Q, CCUS development progress remains sluggish primarily due to significant initial capital outlays (**Figure 1**). Additionally, there is a minimal or even negative rate of return on CCUS projects. A substantial portion of these costs is attributed to the use of costly Corrosion-Resistant Alloys (CRA), crucial for countering the harsh corrosion challenges posed by supercritical CO_2 with impurities in the entire CCUS value chain.

Corrosion issues affect the entire value chain of CCUS, including capture, transportation, and injection processes, all of which have been thoroughly examined. The use of CCUS is limited to materials with exceptionally high corrosion resistance, significantly increasing project costs and posing a major challenge to widespread adoption, especially in developing countries like China and India, which are major contributors to global CO_2 emissions (**Figure 1**). In established Carbon Capture and Storage (CCS) projects in Western countries, expensive high-end corrosion-resistant alloys, such as 25Cr, are widely used [2]. Due to the high costs of material selection, a survey indicates that injecting one ton of CO_2 in a CCS project ranges from $15 to $30, which is 3–6 times higher than the average global living expense of $5 per day [3]. Consequently, China, the top CO_2 emitter, is conservative in investing in CCS projects and focuses on CCUS-EOR (Enhanced Oil Recovery) operations because of the less severe corrosion issues and the economic return from produced fossil fuels, despite the controversial carbon reduction efficiency. As for the second-largest CO_2-emitting developing country, barely any CCUS projects are initiated or planned in India as of 2024. Therefore, reducing material selection costs plays a significant role in promoting the broad adoption of CCUS projects, which is essential for achieving

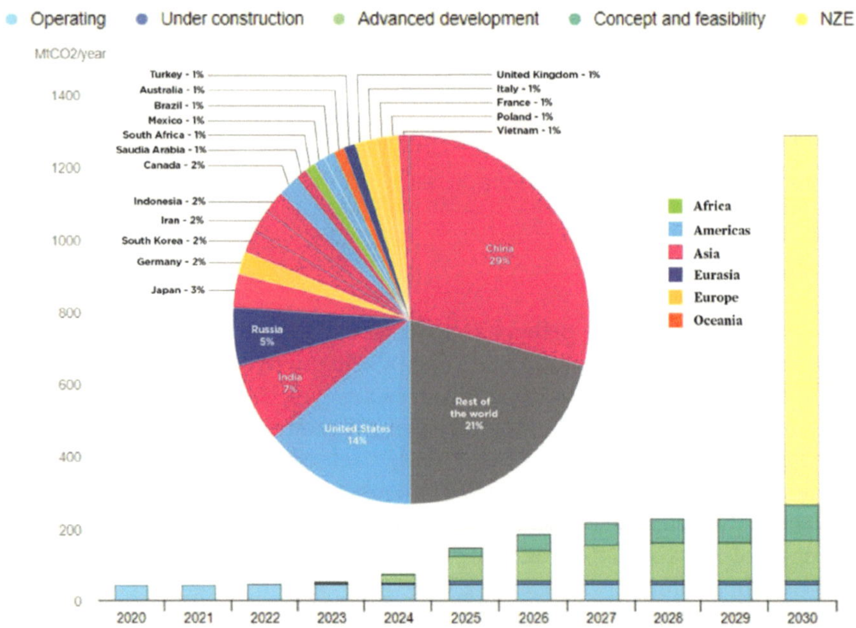

Figure 1.
CCS plants in operation and planned through 2030. Yellow indicates the capacity needed for Net Zero Emissions (NZE). The inset shows the top CO_2-emitting countries from fossil fuels [3].

carbon reduction goals. Fortunately, surface engineering solutions combined with cost-efficient carbon steel have proven to be a highly promising solution to this urgent and challenging issue.

2. Corrosion for metal base

Fundamentally, irrespective of the corrosiveness of CO_2 or associated impurities, or their oxidative nature, they are unlikely to pose any issues within the temperature range of CCUS unless there is the presence of a liquid water phase. Base materials are categorized into carbon steel/low alloy steel and corrosion-resistant alloys (CRA) in downhole tubular applications. Regarding corrosion failure, consider the example of the CCUS injection well tubular. Corrosion can be caused by an extremely high corrosion rate (general corrosion) as well as local corrosion or pitting corrosion, which pose greater challenges. Unlike uniform corrosion that is visibly evident on the surface, local corrosion often manifests as small pinholes. In practical scenarios, the majority of failures in CCUS injection wells are actually attributed to local corrosion, leading to perforation (refer to **Figure 1a**). Beyond corrosion concerns, another crucial factor contributing to failures is Environmentally Assisted Cracking (EAC, as depicted in **Figure 2b**). This encompasses various mechanisms, including sulfide stress cracking (SSC), stress corrosion cracking (SCC), hydrogen-induced cracking (HIC), and others.

The oil and gas industry boasts extensive experience in materials selection for wells, yet the conditions for CO_2 injection wells diverge from those of "conventional" oil and gas wells, primarily due to two distinct reasons. Firstly, the elevated CO_2 content and pressure, reaching several hundred bars, results in an exceptionally high corrosion rate. This phenomenon cannot be simply elucidated by the conventional CO_2 corrosion theory based solely on the low pH value resulting from CO_2 dissolution in water. Noteworthy examples include Cailly et al.'s [4] findings of a steel corrosion rate of 25 mm/y at 65°C and 1 MPa CO_2 pressure, escalating to 250 mm/y at 82°C and 16 MPa CO_2 pressure. Pimenta et al., reporting on onshore CO_2-EOR projects at ADNOC, cite a corrosion rate of 30 mm/year [5]. As a rule, the general corrosion rate is conservatively estimated at 10 mm/year. These figures are notably high, considering the oil and gas industry typically adheres to an acceptable corrosion rate of only 0.025 mm/year, as per industrial standards like NACE RP0775 [6]. To comprehend these heightened corrosion rates, a novel corrosion mechanism specific

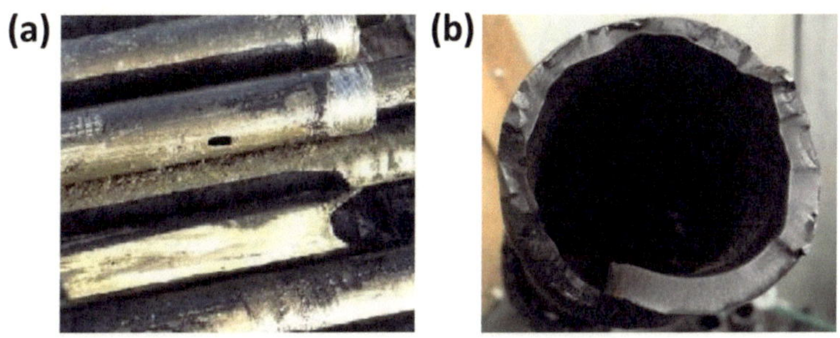

Figure 2.
Photographs illustrating the failure of CCUS injection tubing due to severe pitting (a) and environmental-induced cracking (b).

to supercritical CO_2 corrosion—termed the Direction Reduction mechanism—has been uncovered under these unique conditions. In the traditional theory, when CO_2 dissolves in water, it undergoes a reaction, forming carbonic acid (H_2CO_3, Eq. (1)). This carbonic acid can subsequently dissociate into bicarbonate ions (HCO_3^-) and hydrogen ions (H^+) (Eq. (2)), with bicarbonate ions (HCO_3^-) further dissociating into carbonate ions (CO_3^{2-}) and hydrogen ions (H^+) (Eq. (3)). The attachment of hydrogen ions (H^+) to the steel surface forms a cathode, initiating corrosion (Eq. (4)). Extremely high CO_2 pressure leads to a high concentration of H^+, resulting in a low pH value and an exceptionally high corrosion rate. Under CCUS operational conditions, the injection of CO_2 into the formation water phase could lower its pH value to <3, especially in the absence or limited presence of pH-buffering minerals. This environment is highly acidic compared to the typical pH above 4 in traditional oil and gas production environments. However, actual tests in CCUS conditions have shown that the corrosion rate of steel is much higher than that caused by H^+ or pH value alone. For instance, Nesic et al. [7] noted that the corrosion limit current in aqueous solutions containing CO_2 was higher than that in HCl solutions at the same pH value. To explain these exceptionally high corrosion rates, the direct reduction mechanism has gained broad acceptance. In this theory, steel can be directly oxidized by positive monovalent hydrogens in H_2CO_3 and HCO_3^- on its surface (Eqs. (5) and (6)). Additionally, Gulbransen and Bilkova conducted linear polarization resistance (LPR) and weight loss measurements to study the process of metal corrosion by CO_2, also revealing that the direct reduction of carbonic acid and acetic acid played crucial roles in determining overall corrosion rates [8]. Therefore, more oxidation species, encompassing both dissociated and undissociated carbonic acid, play a role in the corrosion process, leading to an exceptionally high corrosion rate in CCUS conditions:

$$CO_2 + H_2O \rightleftharpoons H_2CO_3(aq) \tag{1}$$

$$H_2CO_3(aq) \rightleftharpoons H^{+(aq)+HCO_3^{-(aq)}} \tag{2}$$

$$HCO_3^{-(aq) \rightleftharpoons H^{+(aq)+CO_3^{2-(aq)}}} \tag{3}$$

$$Fe(s) + 2H^{+ \rightleftharpoons Fe^{2+(aq)+H_2(g)}} \tag{4}$$

$$Fe(s) + 2H_2CO_3(aq) \rightleftharpoons Fe^{2+(aq)+2HCO_3^{-(aq)+H_2(g)}} \tag{5}$$

$$2Fe(s) + 2HCO_3^{-(aq) \rightleftharpoons FeCO_3(s)+H_2(g)} \tag{6}$$

The second reason for divergence lies in the fact that CO_2 is not typically "pure," signifying the potential presence of low concentrations of impurities contingent on the CO_2 source and capture technology employed. While impurities like H_2S and organic acids exist in traditional oil and gas production, those from the captured CO_2 stream, including SO_X and NO_X, possess oxidative characteristics not commonly found in "conventional" oil and gas operations. These oxidative impurities not only contribute to exponentially higher corrosion rates due to their elevated electrochemical potentials but can also induce pitting or cracking failures in high-end Corrosion-Resistant Alloys (CRA) that have reliably functioned in the oil and gas industry for decades.

Corrosion-resistant alloys are engineered materials designed to withstand corrosion and deterioration in aggressive environments by forming a protective oxide layer on their surface when exposed to oxygen or oxidizing agents. This passive layer, primarily composed of chromium oxide (Cr_2O_3), is highly stable and acts as a barrier, preventing further corrosion of the underlying material. The resistance of the alloy to CO_2 attack tends to increase with higher Cr content. To further augment the resistance of Cr-bearing alloys against CO_2 and other sour gases, additional metals such as nickel (Ni), molybdenum (Mo), and titanium (Ti) have been incorporated into the alloys. This stability of this protective layer can be roughly gauged by the Pitting Resistance Equivalent Number (PREN), calculated as %Cr + 3.3 * %Mo + 16 * %N. Typically, higher contents of the costly Cr and Mo elements correspond to better corrosion resistance, albeit at an increased alloy cost. **Figure 3** illustrates the general price ratio of various CRA grades compared to carbon steel, along with their respective PREN numbers. Across the spectrum of 9Cr to 25Cr CRAs, the cost and their associated PREN are closely interrelated [10]. Given the substantial volume of metal in downhole tubulars, minor variations in these grades can translate to millions of dollars in well costs. Based on extensive test results and field experience, 25Cr is generally required for CCS applications in the current project [2]. An even higher grade may be necessary as impurity levels increase when CCUS projects are more broadly adopted. This could lead to nearly a tenfold increase in material selection costs, which is a critical reason why CCUS is difficult to promote globally today and even harder to accept in developing countries and economies.

Therefore, cost-efficient surface engineering technologies, such as coatings, platings, and liners, emerge as effective solutions to prevent severe corrosion issues in carbon steels. These technologies have a proven track record in the traditional oil and gas industry and have undergone extensive development and testing for CCUS applications with preliminary success. However, each technology has its own advantages and limitations, which restrict their application to different ranges of corrosion severity. None of them are currently technically mature enough to replace CRA globally. Nevertheless, their potential is high, and the economic return is evident. To achieve this goal, more research needs to be conducted to address these limitations.

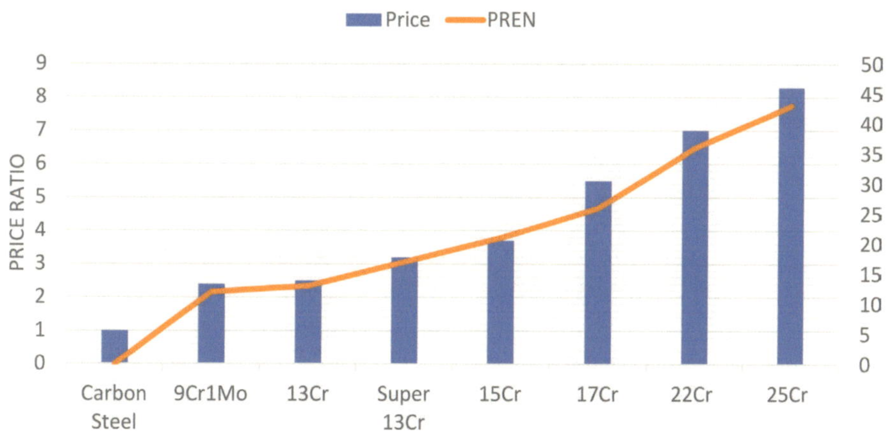

Figure 3.
Overall price ratio comparison between different grades of CRA and carbon steel, along with their corresponding Pitting Resistance Equivalent numbers (PREN) [9].

3. Polymer coatings

Coating represents a conventional method for corrosion prevention, serving as a protective barrier that physically obstructs the interaction between the steel and corrosive agents. This method finds widespread application in various segments of the oil and gas industry, encompassing upstream, midstream, and downstream operations. In essence, to be deemed suitable for challenging CO_2 environments, coatings must exhibit high adhesion, corrosion, and moisture resistance, compatibility with the specific environment, and sufficient structural strength to prevent any form of disbondment.

3.1 Suitable coating material and formulation

To be an effective barrier, a coating material must be chemically stable in the corrosive environment and have low permeability to prevent corrosive media from reaching the base metal. Various coating systems, like epoxy, polyurethane, and polyvinylidene fluoride, are suitable for different environments, each with limits on temperature, acidity, and cost. For example, PE and PP are cost-effective and chemically resistant but are limited to below 80°C. Fluoropolymer coatings resist chemicals well and can exceed 150°C but are costly. Epoxy is balanced but degrades in acidic conditions in combination with H_2S. Each coating grade, differing in molecular structure and filler systems, must be compatible with the targeted fluids, verified through lab tests or vendor specifications.

Effective barrier performance and low permeability are crucial in selecting a coating system. Polymer spaces are typically much larger than corrosive fluid molecules like water and CO_2. All polymers are theoretically permeable to downhole fluids, eventually leading to corrosion and debonding if the coating material remains intact. Additives in polymer formulations, such as stabilizers, pigments, and plasticizers, can leach out, causing polymer degradation. Supercritical CO_2, acting as a solvent, poses a significant risk to internal polymer coatings in wells. Two methods reduce polymer coating permeability: using a highly condensed or cross-linked polymer network (e.g., multifunctional epoxy resin) to minimize spaces between polymer chains and adding inorganic fillers like Al_2O_3, mica, or graphene, which have negligible permeability. Flake-shaped fillers are preferred as they create a tortuous path, increasing barrier effectiveness (**Figure 4a**, [11]). Managing curing chemistry and filler dispersion in real applications requires significant R&D and field experience, distinguishing high-quality coatings from ordinary ones despite similar claims.

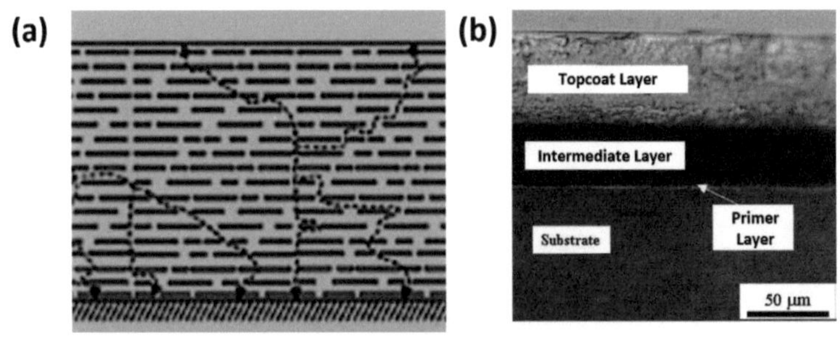

Figure 4.
Illustration depicting the impact of flake filler on minimizing gas permeability (a) and the microstructure of contemporary coatings (b) [11].

Lastly, Rapid Gas Decomposition (RGD) is a common failure in non-metal materials used in downhole environments, such as rubber seals and coatings [12]. It occurs when fluids like water or CO_2 diffuse into the polymer network, causing swelling. These fluids generate internal pressure, and if external pressure drops rapidly, the internal pressure can cause blistering and chemical degradation of the polymer matrix. RGD is influenced by material properties and field operations. To mitigate RGD, materials should have low permeability and low solvent storage capacity, necessitating the selection of polymers with low affinity for solvent molecules. Coatings must be fabricated with high quality and free from defects like voids and cracks. Achieving this requires thorough quality assurance and control (QA/QC), technical expertise, and proper equipment from coating application companies, as well as systematic quality control from end users and third-party inspectors.

3.2 High adhesive strength

Impeccable coating materials are ineffective if bonding with the steel substrate is subpar. Downhole tubulars experience elongation under tensile or bending loads due to high temperatures, self-weight, and wellbore deviation, making poor bonding prone to delamination [13]. This can lead to blistering and peel-off when corrosive fluids penetrate, causing corrosion and further coating failure. No coating is flawless, especially on the inner diameter (ID) of tubulars, where application is challenging due to poor accessibility. Voids or holidays within industrial standards can trigger immediate corrosion, spreading under the coating. Good bonding decelerates this process and extends the coating's lifespan.

Ensuring robust bonding involves selecting appropriate coating materials and effective surface treatment of the base metal. Choosing coatings with strong affinity to metal, like polar epoxy over non-polar PE, is straightforward, but surface treatment is crucial for high bonding strength. Surface treatment often costs more than the coating material itself. To ensure intimate interaction with the metal surface, layers of oxidation and contamination (e.g., grease, dust) must be meticulously removed using solvent baths and sandblasting. Advanced technologies like plasma cleaning are also employed. Achieving a fully cleaned surface is complex and expensive, requiring research into grit material, size distribution, and treatment levels during sandblasting to optimize metal surface profiles for bonding strength. After cleaning, metal surfaces can quickly re-oxidize, requiring precise environmental control and robust project management to apply the coating promptly. Inadequate surface treatment often leads to early coating failure. Therefore, stringent QA/QC measures are essential, with third-party inspectors mandated to scrutinize each coating process step to ensure quality standards are met.

3.3 Good structural integrity

Polymer coatings are generally softer than base metals, and factors like thermal expansion mismatch and internal stress during curing limit their optimal thickness to less than 1 mm [14]. Consequently, these coatings are prone to damage, especially in downhole conditions. Wireline operations, essential for well monitoring and leak detection, pose a significant risk to internally coated tubing during conveyance, as nearly all producing wells undergo periodic wireline work. Despite guidelines from NACE International RP0291-91 to minimize wireline damage, compliance is inconsistent, and wireline damage is often observed during inspections of coated tubing from production wells [15]. Additionally, solid particles like sand in the flow can cause erosion issues.

To effectively meet the requirements of compatibility with corrosive environments, barrier effectiveness, strong adhesion to the base metal, and structural strength to withstand operational stresses, a modern coating typically consists of three layers: Primer, Intermediate, and Topcoat (**Figure 4b**).

- *Primer layer:* Directly in contact with the base metal, this layer is designed for strong adhesion and may contain bonding agents. It has a compact, adhesive microstructure with good wetting properties to ensure robust adhesion to the substrate.

- *Intermediate layer:* Acting as the primary barrier against corrosion, this layer requires a dense network and appropriate fillers to minimize fluid permeability.

- *Topcoat layer:* Positioned as the outermost visible layer exposed to the environment, it provides protection and may include pigments, additives, or other components for color and scratch resistance. If the Intermediate Layer is structurally sufficient, the Topcoat Layer may be optional.

Strong bonding between these layers is critical for their collective functionality. Each layer is typically formulated differently, possibly using distinct base polymers or formulations developed by different companies, which adds complexity to selecting the appropriate coating.

Based on field experience and failure analysis, it is widely believed that coating materials contribute approximately 20% to final performance, while the remaining 80% depends on the application process. Factors such as metal surface preparation, application methods, curing procedures, environmental control, and QA/QC measures are critical. Even with identical coating materials, significant performance variations can occur among different coating application companies. Industrial standards from organizations like the Association of Materials Protection and Performance (AMPP), the American Petroleum Institute (API), and the American Society for Testing and Materials (ASTM) regulate the coating process to ensure reliability. Professional inspectors play a crucial role in verifying compliance with these standards, particularly for downhole tubulars and internal diameter (ID) coatings essential for CCUS injection wells. Specialized tools are necessary due to confined spaces and challenging geometries, impacting cleaning, spray coating, and inspection processes. Despite meeting corrosion protection criteria in lab tests, real-world performance may differ significantly, highlighting the importance of thorough field testing. For instance, while some coatings show low corrosion rates in lab conditions as testing coupons, only a few can reliably protect wells for over 5 years without inhibitors, assuming no physical damage during operation [16]. End users often employ autoclave aging testing to measure corrosion rates to assess coating material effectiveness, but this approach can be extremely misleading. Coupons represent the highest quality a vendor can provide, while the real internal coating of tubing may contain defects due to challenges in coating application and lapses in inspection.

In addition to selecting quality coating materials and employing effective coating applications, the final piece of the puzzle lies with the end user, be it an energy company or injection well operator. They must choose coating grades and processes that are fit-for-purpose, striking a balance between performance and cost. This task can be quite challenging [11]. Finding coatings that consistently perform well is akin to hitting a moving target – the coating remains constant, but operational parameters continually change. Moreover, a coating that excels in one oilfield may be required to be performed in a different oilfield where subtle yet crucial differences in environment exist. Therefore, end users not only need a clear understanding of downhole corrosion environments but also must systematically and scientifically test coating samples to simulate downhole conditions as closely as possible. To qualify internal coatings for downhole tubulars in CCUS applications, autoclave tests are typically necessary to assess both compatibility and bonding of the coating, following the standard testing method NACE-TM0185 [17]. It is worth noting that, in addition to testing flawless coating samples, those with artificial damages must also be tested to evaluate real downhole performance, including potential issues like low-quality coatings (e.g., voids in the coating [18]) and physical damage from operations (e.g., wireline scratches). Rock arm tests (NACE TM0183 [19]) are essential to evaluate the performance of internal coatings under high-pressure, high-temperature (HTHP) conditions. Wireline abrasion tests (NACE RP0291 [20]) and impact damage tests (ASTM D 2794 [21]) are conducted to assess the structural strength of coating materials. After these tests, an overall rating is calculated for each coating, considering every aspect of coating performance. Fit-for-purpose coatings can then be selected based on this comprehensive database.

Based on field experience, coatings have been extensively employed in CCUS applications, yielding differing levels of success [22, 23]. Chevron, for instance, conducted a comprehensive evaluation of its SAROC unit after a decade of operation and reported its findings [16]. In this study, the injection tubing was polymer-coated. Their results indicated that an epoxy-modified phenolic coating proved most successful, with blistering issues observed only when applied excessively thick (>0.17 mm). Powder-applied epoxy demonstrated the highest resistance, boasting an average service life of around 50 months. Unocal, in their Dollarhide Unit, utilized plastic-coated injection tubing. However, corrosion problems arose during field installation [23]. In the Vikor project in Canada during the 1980s, a phenolic epoxy internal coating was chosen for water alternating gas (WAG) injection wells. Despite this selection, Teflon-based dope was reportedly used to address leakage issues at the threaded connections [24]. EnCana, managing the Weyburn CO_2 flood (WAG) in Canada, initially employed alloy 625 clad injection X-mas trees. However, due to cost considerations, they transitioned to internal coatings, many of which proved unsuccessful [25]. In China, various fields have reported the use of polymer coatings in CO_2-EOR injection wells. However, field experiences indicate that these coatings can hardly survive beyond 5 years in the best-case scenario, despite lab testing suggesting longer durability. Field cases have shown that the lifespan of coatings is limited, and their performance is not consistently reliable. As in the oil and gas industry, it is generally believed that coatings are not a reliable choice for corrosion protection if the corrosion rate of the base metal exceeds 2 mm/year under given conditions [14]. As discussed earlier, the CCUS CO_2 corrosion rate far exceeds this threshold, reaching at least 10s of mm/year, even without the presence of oxidizing impurities. Consequently, polymer coating is not recommended for CCUS injection wells, unless the project lifespan is short, or the operator prefers frequent tubing replacements.

In summary, existing commercial polymer coatings lack the mechanical robustness needed to fully withstand downhole damage caused by tool impact or wireline wear. Additionally, they may not provide the required quality for a 100% barrier effect. Successful applications of coatings for corrosion prevention necessitate continuous chemical inhibition, effectively addressing coating defects where the base metal is exposed. Without chemical inhibition, coatings might reduce the frequency of corrosion failures but will not extend the time to failure. The combination of coating and inhibitor has proven effective in traditional oil-producing wells, showcasing synergistic effects [26]. For instance, in the Philips Ekofisk wells, the coating alone protected for only 19 months in production wells, while batch treatment with a corrosion inhibitor every 30 days extended tubing life to 7 years [27]. Lab tests also indicate that a corrosion inhibitor can significantly enhance the lifetime of coatings. Unfortunately, to the best of the author's knowledge, corrosion inhibitors suitable for transportation by supercritical CO_2 have not been reported [28]. Additionally, batch treatment of inhibitors is challenging and economically prohibitive due to the extremely high pressure in injection wells. Therefore, specific inhibitors designed for CO_2 injection wells are currently under intensive investigation. Until such inhibitors are well-developed, polymer coating remains an unattractive option for CCUS injection well applications. It may still find utility in EOR projects where operations are of short duration, lasting only a few years.

4. Metal plating

Numerous organic anti-corrosion coatings face limitations in deployment under supercritical CO_2 ($ScCO_2$) conditions due to the exceptional solvent properties of $ScCO_2$ and the vulnerability of polymer coatings to physical damage. Utilizing plating, a metallic coating with elevated hardness and wear resistance, could offer a more robust alternative to weak polymer coatings [29]. The low permeability of inorganic materials addresses blistering concerns, extending the pressure up-limit. Additionally, the high hardness and strength of plated metals significantly enhance resistance to physical damage during operations (e.g., wireline activities). The plasticity of plated metals and their strong bonding with the base metal ensure coating integrity even under tubular deformation, providing a notable advantage over other inorganic coatings like ceramic coatings and surface nitriding, which may be too brittle for practical applications given their anti-corrosion performance.

Plating processes typically involve chemicals, including potentially hazardous substances like heavy metals, cyanides, and strong acids. Improper handling of these chemicals poses risks to the environment and human health. Wastewater containing these pollutants requires proper treatment before discharge, leading to stringent regulations in Western countries and limited large-scale manufacturing of plated tubulars in these regions. Consequently, field experience with internally plated tubing, particularly in CCUS injection wells, is limited. Early testing of plated tubing in CO_2-EOR injection showed some success [16], but concerns over galvanic corrosion with the base metal in the presence of defects led to the abandonment of this approach. In contrast, the plating industry in China has developed extensively due to less stringent environmental regulations. Large-scale facilities plate tubular structures with metals such as Ni-P and Ni-W. The cost is slightly higher than that of conventional polymer coatings but significantly lower than corrosion-resistant alloy (CRA) tubulars, driving research and application efforts primarily by Chinese institutes and oilfield operators.

In contrast to widely used chrome plating, known for its high corrosion resistance and hardness, Ni-P (nickel-phosphorus) plating is the primary plating material extensively applied to coat the internal surfaces of tubing [29]. This choice is attributed to its initial low cost and environmental friendliness during the early stages of the CCUS campaign. Distinguishing itself from other metal coatings with clear grain boundaries (which are susceptible to corrosion and fluid penetration), Ni-P exhibits an amorphous microstructure [30], resulting in a dense coating with an exceptional barrier effect. Depending on the formulation, Ni-P plating can achieve hardness levels of up to 70 HRC, potentially exceeding that of most steel used in downhole tools like wireline. This theoretically makes Ni-P plating less susceptible to mechanical damage than polymer coatings. Additionally, the metal-metal bonding in plating is typically stronger than polymer-metal bonding, endowing plating with higher bonding strength than polymer coatings. In reports by Zhu et al., Ni-P plating is compared to several corrosion-resistant alloys commonly used in downhole operations [29], including 13Cr-L80, 28Cr-L80, 316 L stainless steel, and Inconel 625, with Ni-P's cost being significantly lower. As a result, Ni-P plating has found widespread use in Chinese oil fields [31], encompassing production wells, water injection wells, pipelines, and even CO_2 injection wells where polymer coating falls short in providing sufficient protection and expensive corrosion-resistant alloys are not economically viable. However, there are varying reports on its performance—some consider it a highly successful technology, while others report poor performance and express reluctance to use it. Failure analysis indicates that not all Ni-P plating is manufactured to the same quality standards, and significant differences in performance are contingent on the plating process and the rigor of quality assurance and quality control (QA/QC) measures. These differences can be attributed to the following factors essential for achieving effective plating:

- *Achieving defect-free plating and addressing the Galvanic Corrosion Effect is a formidable challenge.* As previously discussed, defects such as holidays, voids, and scratches are nearly unavoidable in polymer coatings during both fabrication and operation. Despite the inherent robustness and resistance of these coatings to physical damage, rectifying such defects in the plating process poses significant difficulties. These defects can initiate corrosion, leading to premature failure in polymer coatings. However, the repercussions are even more severe in plating due to the occurrence of galvanic corrosion. For effective corrosion resistance, the plating material must be more noble than the base metal. Consequently, in instances where defects expose the base metal, the formation of a galvanic cell becomes highly probable. This scenario represents a critical issue, as it results in a precarious cathodic (plating) to anodic (defect) area ratio, facilitating rapid corrosion specifically within the defect area. The corrosion process can swiftly penetrate through tubing wall, despite the majority of the tube remaining structurally intact. This failure mechanism has been extensively documented, particularly in the Chinese field.

- *Achieving high-quality plating presents a significant challenge in surface treatment.* As discussed in the context of polymer coating, effective surface treatment is essential to establish a coating with robust bond strength, necessitating the removal of surface contamination and oxidation layers to expose the base metal for subsequent coating. This imperative holds true for plating, with an added layer of complexity—the surface must not only be clean but also smooth. While sandblasting is a typical method for achieving a desired surface profile in polymer coatings, creating a rough surface that enhances bonding through increased surface area and a physical locking effect, this same surface profile proves challenging for the electroplating process. Electroplating relies on the electric current's flow through the electrolyte to deposit metal ions onto the substrate. A rough or uneven surface introduces the risk of non-uniform current distribution, where areas with high points or roughness receive higher current density, leading to thicker plating, while low points may receive insufficient current, resulting in thinner or no plating at all. Given the thin nature of most platings (typically around 20 micrometers) and the surface roughness following sandblasting (which can reach up to 10 micrometers), achieving uniform plating without defects becomes inherently challenging. Consequently, a polishing procedure is essential after sandblasting to render the inner diameter (ID) of the tubing exceptionally smooth. However, due to the limited accessibility of the tubing ID, this polishing process becomes both expensive and challenging to monitor. This, in turn, significantly increases the plating cost and is frequently omitted by low-quality vendors.

Therefore, the performance of Ni-P plating is heavily reliant on processing and quality control. Unfortunately, there is no international industrial standard governing these products, emphasizing the importance of choosing a qualified vendor for end users. While the superiority of plating over polymer coating is predicated on the strength and wear resistance of metal plating, this assertion is subject to scrutiny given the relatively thin nature of such plating. Observing market trends, Ni-P plating is gradually losing ground due to increasingly frequent reported failures. To address this challenge, the industry is shifting toward Ni-W coating, primarily composed of nickel and tungsten elements [32]. Pictures of Ni-W internal coated tubing are shown in **Figure 5**. Ni-W exhibits greater strength and thickness compared to Ni-P. It is more chemically stable and harder, with thicknesses reaching almost 200 μm and beyond [33]. This increased thickness provides better tolerance in surface treatment,

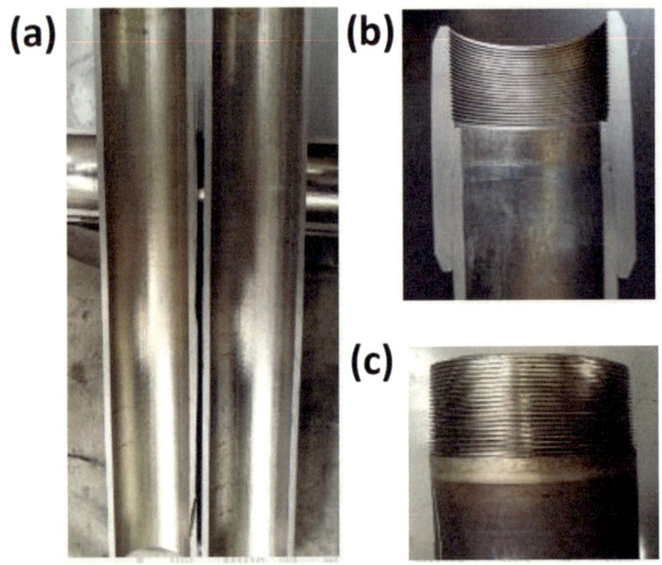

Figure 5.
Photographs depicting carbon steel tubing with Ni-W plating on the internal surface, captured in inner diameter (ID) view (a), as well as in the thread regions (b and c) [9].

potentially justifying the elimination of the expensive polishing procedure, although its preference persists. Field experiences in the oilfields of China report superior overall properties compared to Ni-P coating, showcasing excellent wear resistance, improved corrosion resistance in acidic environments compared to 18Cr-8Ni steel, and commendable thermal stability. Ni-W coatings have been found to be extensively used in CO_2-EOR injection wells and production wells in the region [34]. Despite being two to four times more expensive than Ni-P plating, depending on plating thickness, it still proves more cost-effective than Corrosion-Resistant Alloys (CRA) grades required in similar corrosion conditions.

Currently, the application of Ni-W is in its early stages, with an oil field company reporting successful operation in CO_2-EOR wells for over 7 years without the use of inhibitors. This technology holds promise for Enhanced Oil Recovery (EOR) applications. However, it does not fully address concerns regarding severe galvanic corrosion. Laboratory tests reveal that pitting corrosion in defects can exceed 4 mm/year under relatively moderate corrosion conditions [32]. Successful applications in corrosive CO_2 injection wells depend on assuming zero defects, which is relatively easier for Ni-W compared to Ni-P due to its greater thickness. Nevertheless, there is currently no effective Quality Assurance/Quality Control (QA/QC) tool or testing method to guarantee this assumption. The plating of the threaded area poses an additional challenge, as it can be easily damaged if the plated tubing is not assembled following strict rules and is subjected to unexpected loads or vibrations. As of now, the threaded area remains a weak point, with over 80% of reported failures occurring in this region. Similar to polymer coating, the use of corrosion inhibitors is strongly recommended due to synergistic effects, as supported by field tests [35]. However, since this technology is still being optimized, Ni-W plating could potentially revolutionize CO_2-EOR wells, including CCUS-EOR wells. Further tests are necessary to evaluate the effectiveness of oxidizing impurities in anthropogenic CO_2 resources. Until these data are thoroughly tested, Ni-W is not recommended for long-term requirements in CCS injection wells. Nonetheless, exploring even thicker metal protection layers, such as bimetallic tubes or metal cladding, holds promise, as discussed in subsequent sections.

5. Polymer composite liner

As previously discussed, coatings, primarily thermoset-based, were thin, lacked substantial impact resistance, and exhibited limited flexibility. This often led to breaches in the barrier, leaving areas of the tubing vulnerable to mechanical damage during handling and in-service operations, ultimately resulting in tubing leaks. However, there is an inherent upper limit (typically less than 1 mm) for polymer coating thickness, beyond which coatings tend to delaminate during application due to internal stress. To address these challenges, a preferable solution involves the use of a prefabricated polymer liner inserted inside the tubing. These liners boast significant thickness (typically 2–6 mm, depending on design). Given the brittle nature of thermosets (typically <10% elongation), thermoplastic polymers emerge as an ideal choice due to their flexibility (up to 30% elongation). The thermoplastic polymer liner, slightly larger in outer diameter than the steel tube's inner diameter, is pulled through the steel tubing in a deformed shape to eliminate gaps between them. Additionally, thermoplastic polymers can be heat-treated and shaped into a flange at the tube end (see **Figure 6a**), ensuring a continuous barrier post-installation.

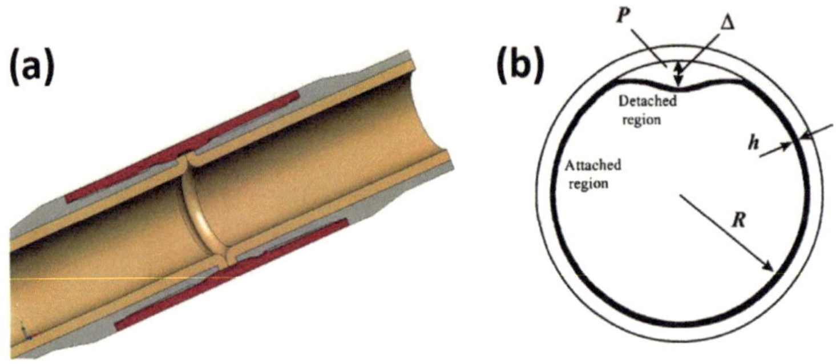

Figure 6.
Diagrammatic representation of tubular structures with thermoplastic lining (a) and depiction of the rapid gas degradation failure mechanism (b) [36].

Moreover, the increased ductility of thermoplastics makes them highly resistant to wear and impact, offering protection against abrasion from sand, wirelines, rotating and reciprocating sucker rods, and coiled tubing. This durability extends to flexing or yielding, which is common in the thin pin ends of connection systems for downhole tubulars. Various thermoplastic grades, with different cost ranges and chemical/thermal stability, have been employed in downhole conditions with varying degrees of corrosion severity. For instance, polyolefin is typically used up to 99°C, Polyphenylene Sulfide (PPS) can handle temperatures as high as 175°C, and (Polyether Ether Ketone) PEEK is utilized for extreme conditions up to 260°C [36]. Consequently, thermoplastic liners have demonstrated considerable success in traditional oil and gas wells, particularly where wear resistance is crucial, such as in tubing ID protection from sucker rods in reciprocating rod pump systems. They have also found widespread use in water injection wells. However, it is worth noting that this technology is not suitable for gas injection wells (including CO_2) due to the severe rapid gas degradation (RGD) effect. In detail, the injected gas permeates the polymer, accumulating between the liner and tubing ID, causing bonding destruction and generating gas pressure. This accumulated gas could collapse the entire liner when the injection pressure drops (a frequent occurrence due to pressure fluctuations), as illustrated in **Figure 6b**. These failures have been widely observed in CO_2-EOR wells in both the US and China, leading to the abandonment of this approach in CCUS injection wells.

To mitigate the RGD issue, a more effective solution involves the use of a rigid and robust polymer composite liner. The principle of the composite material lined carbon steel tubing system is illustrated in **Figure 7**. The liner consists of a lengthy, slender-walled pipe made of fiber-reinforced polymer (FRP), typically utilizing glass fiber epoxy (GRE) material [37]. GRE tubular structures find extensive use as a corrosion-resistant alternative for fluid transport across various industries, with epoxy resins being the preferred material choice as the polymer matrix. These GRE liners are inserted into the steel tubing and secured in place by filling the annulus with a cement layer. The ends of the FRP liner are safeguarded by compression-molded glass fiber-reinforced flanges (known as flares), providing mechanical protection against well operations. Couplings are shielded by a corrosion barrier ring. This arrangement is adaptable to various coupling and connector types, encompassing both standard API and premium (gas-tight connections). This technology, refined since the 1950s, is well-established, and the API RP 15CLT standard has been instituted to regulate both the manufacturing and testing of this composite liner technology [38].

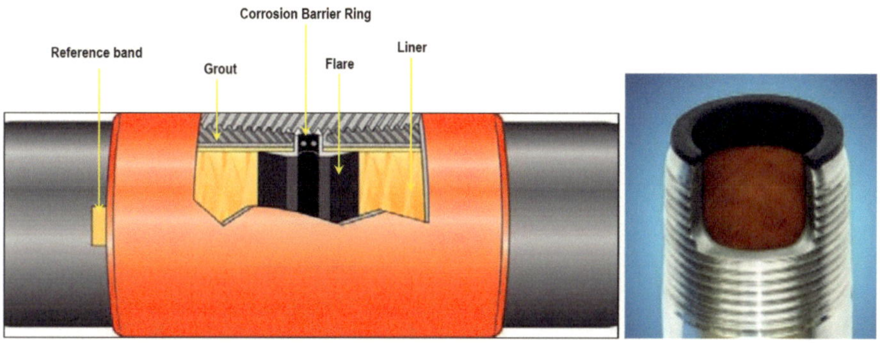

Figure 7.
Schematic illustration of GRE-lined tubular structure (left) and picture of thread area (right) [37].

The primary operational requirement for the liner is to serve as a corrosion barrier for corrosive fluids throughout the tubing string's lifespan. Mechanical loads acting on the liner are transmitted to the steel host tube through a cement grout injected into the annulus space between the liner and tubing. Additionally, the liner is securely anchored to the steel pipe through shear friction at the interfaces of the steel surface, cement, and composite surfaces. The composite's high stiffness and strength prevent liner collapse during pressure fluctuations, maintaining the liner's original shape and ensuring tight contact with outer layers. The liner material must exhibit high chemical stability to prevent material degradation, which could lead to loss of material from the inner surface or material blistering. Various grades of liner materials, contingent on epoxy and fiber composition, are commercially available to address diverse corrosive environments, including CCUS. In comparison to polymer coatings, the liner offers an extended lifespan (25–50 years) [39], resistance to ordinary wireline operations, and no collapse issues under typical operational conditions. Moreover, the densely cross-linked epoxy network in the GRE liner results in a CO_2 molecule permeability that is one million times smaller than that of traditional HDPE liners. This not only diminishes the risk of RGD issues but also enhances the barrier effect significantly. Saudi Aramco conducted aging tests on GRE liners in gas well conditions, predicting a lifetime of over 20 years at 110°C in sweet gas wells, presenting a cost-effective alternative to more expensive 22/25Cr stainless steel [40]. Field experience from Saudi Aramco demonstrates that fiberglass-lined tubing is a low "life cycle cost" solution, with no workovers required since installation. Consequently, fiberglass-lined tubing is applied in Saudi Aramco's high water-cut oil producers, water injectors, and combined water source and injection wells. Statoil reported a similar test, anticipating a 25-year lifetime at 110°C under sour conditions [41]. While the initial capital expenditure (CAPEX) cost of GRE liners surpasses that of coatings, the lifecycle cost (CAPEX plus OPEX) is considerably lower based on ISO 15663 Lifecycle Cost (LCC) analysis principles endorsed by various institutions [42]. This is attributed to GRE liners operating maintenance-free for over 25 years, in stark contrast to coated tubing, which typically requires replacement every 5 years. The cost savings encompass both the tubing (and coating) itself and the labor and rig costs incurred during each replacement [43]. A schematic illustration (**Figure 8**) from Vallourec underscores that GRE has the lowest cost compared to inhibitors, coatings, and Corrosion-Resistant Alloys (CRA) for projects exceeding 6 years [44].

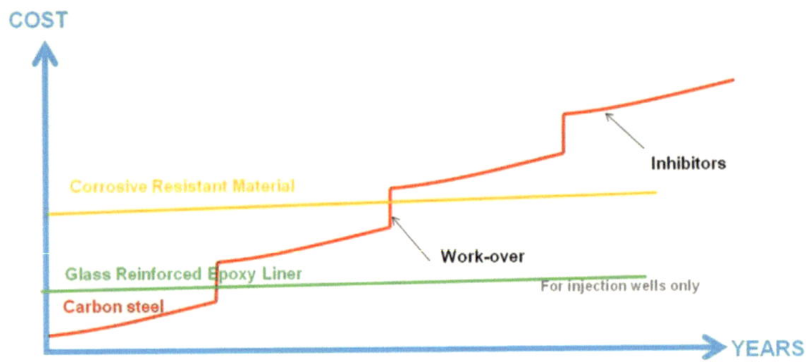

Figure 8.
Comparison of the total cost over time for corrosion prevention in carbon steel, GRE liner, and CRA (Corrosion-Resistant Alloy) [44].

This technology has been extensively utilized in CO_2-EOR injection wells over the past 50 years, boasting a proven track record. It stands out as the optimal choice alongside Corrosion-Resistant Alloys (CRA) materials. Notably, between 1980 and 2018, major players such as Statoil, Chevron, and ExxonMobil deployed over 20 million feet of GRE-lined tubing in CO_2 injection and sequestration projects worldwide. BP has reported the use of GRE liners to replace 25Cr in their water injection wells [45]. Statoil uses GRE-lined tubing extensively in their North Sea WAG wells [46]. In the Permian basin, Oxy has reported broad usage of 6 million meters of GRE liner in their CO_2-EOR injection wells over 20 years [47]. In Canada, this technology has been reported to be broadly utilized in the CO_2-EOR as early as 1986 [48]. In China, the Shengli oil field's report indicates that the performance of GRE liners surpasses that of Ni-P plating discussed in the previous section, albeit at a cost of approximately 70% of that of steel tubing [49]. ISO 17348, a widely accepted standard for materials selection in CO_2-EOR applications, clearly stipulates that only corrosion-resistant alloys (CRA) and GRE liners are acceptable for CO_2-EOR injection wells.

Besides CO_2-EOR or CCS-EOR, GRE liner has been reported to be used in CCS projects designed for long-term operation. It has been reported to be used in the Wabamun Area CO_2 Sequestration Project (WASP) [50]. As a matter of fact, in the EPA's Federal official guidance for Class VI wells (CCS wells) [51], GRE lining is well accounted for as a corrosion-resistant material under article 2.4.2 Corrosion Considerations.

However, despite the wealth of reports, there is limited research or knowledge concerning the impact of oxidizing impurities in anthropogenic CO_2 resources on the aging of GRE liners. Consequently, caution is advised when selecting this technology for CCUS-EOR and CCS projects where the impurity level is extremely high in the injected CO_2 stream. Additionally, due to polymer-related degradation concerns, questions persist about the potential longevity of this technology over a 50- to 100-year operational period if the GRE liner is left downhole during the post-closure stage.

While the GRE liner technology has demonstrated a low lifecycle cost and a successful track record in CCUS injection wells, it is not without its imperfections. Some operators opt for more expensive corrosion-resistant alloys (CRA) instead of this technology due to specific issues, despite its outstanding corrosion resistance performance [52]. The following aspects warrant attention when considering its selection:

- *Diminished inner diameter of the tubing and limitations on the deployment of production logging tools*. The presence of the liner (2–4 mm thick) and the cement filler layer (approximately 2 mm) results in a reduction of the inner diameter of the lined tubular by 4–8 mm. This reduction has the potential to constrain the flow rate of fluids passing through, particularly in smaller tubulars. However, it has been suggested that the reduced friction between fluids and the "slippery" GRE liner could offset this effect to some extent. Given the reduced inner diameter, the feasibility of passing logging tools needs to be carefully assessed.

- *Material selection for GRE requires thorough testing and qualification*. Epoxy is susceptible to degradation by hydrogen sulfide (H_2S), and caution is advised when employing GRE liners in environments with significant H_2S content, either in the CO_2 stream or the formation [53]. In comparison to metals or pure polymers, Quality Assurance/Quality Control (QA/QC) for polymer composites has consistently posed challenges due to the intricate microstructure and the manufacturing difference between the QC testing coupons and the GRE liner itself. Poor bonding between fibers and the epoxy matrix is not uncommon, creating pathways for fluid infiltration (fluid absorption reaching 2–10 wt.%), leading to chemical degradation of both the epoxy resin and its bonding with fibers. Failure in either of these aspects can result in substantial loss of mechanical properties and liner failure well before the corrosion barrier effect is compromised.

- *Restricted working temperature range*. While this may be less of a concern for CO_2 injection wells (typically <90°C), it still necessitates consideration, especially in the case of elevated downhole formation temperatures (i.e., those above 150°C). Various grades of GRE liner products are available, contingent on the grade and molecular structure of epoxy resins, with temperature ratings spanning from 60 to 120°C or higher [26]. The appropriate product should be selected based on thorough testing reports and field experience.

- *Limited pressure rating*. Field experience suggests that fiberglass and fiberglass-lined tubing are not employed when the pressure exceeds 340 bar (34 MPa) [23]. It is important to note that this limit applies to the commodity grade of composite products, and this pressure rating is generally sufficient for the majority of injection wells. There are also premium grades available that can safely be used beyond this limit.

- *Caution is still necessary to prevent damage*. Despite exhibiting superior mechanical wear and resistance properties compared to plastic or epoxy coatings, GRE lining is inherently softer and more brittle than metals. Therefore, wireline operations necessitate specific procedures and precautionary measures to ensure an extended lifespan.

- *RGD remains a concern*. As supercritical carbon dioxide ($ScCO_2$) has the ability to permeate the bulk material of the liner, some trapped $ScCO_2$ gas may be present within the various layers of the composite (e.g., defects) or in the annular space between the composite liner and the carbon steel pipe. The presence of this trapped gas can result in the formation of blisters within the composite layers or even lead to liner collapse (exposing fibers) in the event of rapid gas decompression. Both occurrences can contribute to a reduction in the mechanical properties of the liner and an increase in permeation rates through the material thickness. Fortunately, laboratory tests and field experience indicate that there should be no issues if the depressurization rate is controlled below 1000 psi/min, a rate easily manageable in the field operation.

- *Limited flexibility*. In contrast to steel or corrosion-resistant alloys (CRA) tubulars, GRE-lined tubes are more rigid due to the presence of a rigid composite and cement layer. This rigidity constrains the ability of the lined tube to bend, making it less suitable for highly deviated wells. Well designers are obligated to adhere to the product's operation manual in such scenarios. It also requires special caution during the handling and shipping of the lined pipe.

- *Additional constraints*. The assembly of GRE liners requires expertise and well-trained field engineers, posing potential challenges for quality control and supply chain management. Furthermore, GRE-lined tubing does not offer protection against scale formation from $BaSO_4$ and $SrSO_4$ [54].

6. CRA liners-lined tube and clad tube

In comparing the performance of polymer coating and polymer composite liners, it becomes evident that increased liner thickness significantly enhances anti-corrosion resistance. Consequently, as an upgrade from metallic coating, a metallic liner with thickness in the millimeter range is poised to address issues such as improved wear resistance, complete elimination of voids/holidays, etc. This approach holds an advantage over polymer composite liners due to their low permeability (gas permeability in metal being lower than in polymer) and higher temperature rating. Corrosion-resistant alloys (CRAs) emerge as the optimal choice for liner materials. The use of CRAs to control corrosion in oil and gas systems has a proven track record. Production systems constructed with appropriately selected CRAs, based on rigorous laboratory testing or past field experience in similar environments, have provided a safe, leak-free system throughout the entire project duration. In both Carbon Capture and Storage (CCS) or CO_2-EOR projects, various fit-for-purpose grades of CRA materials, including 13Cr, Super 13Cr, 22Cr, 25Cr, and 718, have undergone extensive testing and widespread utilization [55]. However, the capital expenditure (CAPEX) cost of CRAs is notably high, attributed to the inclusion of expensive alloying elements, particularly for more highly alloyed materials required in corrosive sour production systems. Consequently, CRA liners can be viewed as composite products developed to enable effective and economical utilization of expensive CRA materials. The CRA liner that comes into contact with corrosive fluids is crafted from a corrosion-resistant alloy, while the less expensive backing steel provides the strength and toughness necessary to maintain mechanical integrity. Utilizing high-strength backing steel allows for a reduction in wall thickness compared to solid CRAs, thereby decreasing fabrication time and costs.

CRA-lined tubulars go by various names, such as bimetallic tubes, lined tubes, clad tubes, etc., and they can be fabricated using different technologies that result in significantly varied microstructures and physical/chemical properties. A comprehension of the manufacturing process and structure is crucial for material selection. Broadly speaking, CRA-lined tubular can be categorized into two fundamental types: metallurgically bonded and mechanically lined [56].

6.1 Mechanically lined tubular

In its simplest form, a standard pipe can be internally lined by introducing a seamless or welded liner made of Corrosion-Resistant Alloy (**Figure 9a**). If no heat is applied, the liner is hydraulically or mechanically expanded inside the outer pipe. The liner is then secured in place by the mechanical forces of the shrink-fit without forming an integral bond along the full length. Alternatively, the outer pipe may be heated before inserting the liner. Simultaneously with thermal shrinkage, the liner is hydraulically expanded outward against the steel shell. Seal welding is applied at the pipe ends to prevent moisture ingress. Consequently, this manufacturing method does not establish a metallurgical bond between the outer pipe, which provides structural integrity, and the inner liner pipe selected for corrosion resistance. This is the most economical method to fabricate lined tubulars and is widely available in the market. However, this technology is unsuitable for downhole tubular applications, especially for CO_2 injection wells [57]. Firstly, downhole tubulars typically require certain bending, and under such conditions, this technology is inappropriate due to liner buckling resulting from a lack of bonding between the liner and the base steel. Secondly, there is no protection in the thread area. This technology is designed for applications where tubulars are connected by welding, allowing liners to be welded together to provide continuous protection. However, in downhole tubulars, only threaded connections are feasible, and over 80% of corrosion issues originate from there. Although some companies attempt to provide designs to protect the thread area, their reliability cannot be proven. Given that galvanic corrosion can occur between noble CRA and the steel base pipe, severe corrosion is observed when there is failure in these threaded sections. Therefore, the application of this technology is rarely reported, and it is not recommended for any CCUS injection wells.

6.2 Metallurgically bonded tubulars

Metallurgically bonded tubulars are characterized by a metallurgical bond between the structural outer pipe and the corrosion-resistant inner pipe (**Figure 9b**). This type of tubular tends to incur higher costs compared to mechanically lined ones. The most widely utilized method in the oil and gas industry for producing metallurgically bonded tubulars is Clad Plate Forming. Clad plates can be manufactured through three common methods: hot roll bonding, explosive bonding, and weld overlay. These clad plates find extensive use in various processing vessels, separators, heat exchangers, and plates. Once the clad plate is produced, it is shaped into a tubular shell and longitudinally welded along its entire length [58]. This tubular is suitable for pipelines or downstream facilities but is not appropriate for downhole tubulars, as the extreme

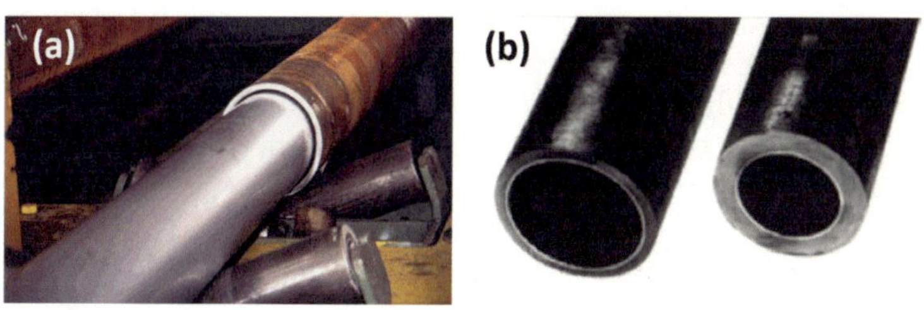

Figure 9.
Photo image of mechanically lined tube (a) and Explosively Bonded tubular (b) [56].

high-pressure necessitates the use of seamless tubing. To create a seamlessly bonded CRA liner tube, there are typically three manufacturing methods:

- *Extruded/pipe mill products:* This type of pipe is manufactured by creating a composite billet with a Corrosion-Resistant Alloy (CRA) pipe nested inside a steel pipe [57]. The two components may be welded together or partially brazed. The composite billet can undergo processing through standard pipe mills such as a plug mill, mandrel mill, or extrusion press. Achieving optimal temperature control of the metal and managing the deformation rate requires considerable skill, as the cladding alloy and backing steel exhibit differences in hot strength. Seamless pipes produced by these manufacturing routes are more susceptible to issues such as lack of concentricity, variations in cladding layer thickness, and wider tolerances on circularity.

- *Explosively bonded tubular:* Explosive bonding involves using a brief, high-energy impulse generated by an explosion to force two metal surfaces together [56]. The explosion effectively removes surface oxides and forms a metallic bond between the carbon alloy backing steel and the Corrosion-Resistant Alloy (CRA) layer. While the two surfaces do not collide instantaneously, they progressively do so over the interface area. This process is generally acknowledged to be labor-intensive and may not be economically viable for long tubular.

- *Weld overlaying:* This process involves depositing a layer of CRA material directly onto the surface of a workpiece, also referred to as "cladding" or "hard facing." It is typically accomplished through methods like hot spray, Gas Tungsten Arc Welding (GTAW), and Gas Metal Arc Welding (GMAW). The choice of welding method depends on factors such as access, welding position, alloy type, dilution requirements, and economic considerations. Due to limited accessibility to the inner diameter (ID) of long tubing, cladding the tubing ID with this technology can be expensive. Quality control is essential to produce a liner without through porosity; otherwise, similar galvanic interactions as with plating material may occur. In some cases, sealing material may need to be applied to fill any potential pores.

- *Centricast pipe:* A distinctive approach to produce seamless clad pipe involves horizontal centrifugal casting technology [57]. In this process, well-refined molten steel is poured into a rotating metal mold with flux. After casting, the temperature of the outer shell is carefully monitored. Once the outer shell has solidified to a suitable temperature, molten Corrosion-Resistant Alloy (CRA) is introduced. The choice of flux, the temperature of the outer shell during the introduction of molten CRA, and the pouring temperature of the CRA are precisely controlled to achieve a robust metallurgical bond with minimal mixing at the interface. Although such pipes have been manufactured and utilized in oil and gas services, they are not currently produced, primarily due to their high cost.

The cost savings derived from using liners instead of solid corrosion-resistant alloy (CRA) are particularly significant when the total thickness increases or when the cladding grade becomes more intricate and, consequently, more expensive [56]. In fact, clad steel plates have been successfully employed in various applications, including processing vessels, heat exchangers, tanks, material handling and storage facilities, as well as in the production of longitudinally welded clad pipes. For instance, there exists well-established knowledge and standards for lined or clad pipelines, including those designed for CO_2 transportation [59]. In the past decade, several manufacturing methods have been developed for producing CRA clad/lined tubulars. The American Petroleum Institute (API) has also formulated a specification for CRA clad and lined line pipes (API 5LD14) [60]. Clad and lined tubulars have achieved relatively broad acceptance for transporting petroleum products and in refinery applications.

The adoption of these downhole tubulars has not found favor in downhole oil and gas operations (including CCUS), mainly due to the absence of a threaded connection that has demonstrated satisfactory performance [57]. Despite various reported designs, there is still no universally accepted thread or connection design that ensures a consistently reliable CRA protection in the pipe joining area, which has consistently been the failure point of this technology [61]. Additionally, from an economic standpoint, it appears that clad tubing can only effectively compete with solid alloys in large sizes, particularly when there is a significant need for high-grade CRA. This does not align with the requirements of CCUS injection well tubing, the OD of which is typically below 5 inches [56]. For instance, an assessment of plasma transfer arc weld-coated tubular was conducted in 1995. The tubing specimens were coated with an alloy composition similar to alloy C276, and while corrosion properties were systematically investigated, this technology has not progressed to the production of full-size tubing [26]. However, in theory, cladded liners should be a viable option for both CO_2-EOR and CCS, but the manufacturing costs need to be reduced before widespread application can be realized.

7. Conclusion

The material used in injection wells is predominantly in the form of tubulars. Even slight changes in the material selection for these tubulars can significantly impact the overall project cost, making it a key research focus in CCUS injection well applications. Despite being utilized in traditional CO_2 EOR and some CCS-EOR projects, bare steel is only effective when the injected CO_2 stream is pure and can be reliably isolated from free water, such as condensed water or formation water. However, this scenario is unlikely due to the shift toward anthropogenic CO_2 resources and CO_2 storage instead of EOR. The corrosion rate of bare steel would be prohibitively high (exceeding 10 mm/year) for use in modern CCUS injection wells. Polymer coating can offer a barrier effect, but its lifespan is challenging to extend beyond 5 years in harsh supercritical CO_2 conditions. Achieving high quality in the tubing ID poses a significant challenge. Most importantly, thin polymer coatings are highly susceptible to various physical damages during transportation, assembly, and wellbore intervention, making them unsuitable for CCUS applications. The only exception is in projects with a short lifespan and low completion string replacement costs. Although this approach is not aligned with the evolving trends in CCUS, it might still be viable for developing economies initiating small projects to stimulate infrastructure investment.

Plating, involving a layer of noble metal coating with superior corrosion resistance and high strength to withstand physical damage, represents a notable improvement over polymer coatings. This technology has become more attractive, considering its cost is not significantly higher than that of conventional polymer coatings, thanks to well-developed facilities and scale in China. Field experiences, especially with Ni-W plating, have shown promise. However, the risk of galvanic cell corrosion between the plating material and base steel at defective areas necessitates careful consideration. This technology is viable only when perfect plating (100% coverage, no holidays or voids) can be guaranteed. Several inspection technologies in Quality Assurance/ Quality Control (QA/QC) exist to achieve this, but none can provide a 100% guarantee. Currently, there is no industrial standard regulating this, and further developments are required to promote this technology globally in all three CCUS projects.

In addition to CRA, Glass Reinforced Epoxy (GRE) liners have emerged as the primary choice for traditional CO_2-EOR, benefiting from continuous optimization over the past 50 years and offering lower costs. These liners are endorsed by ISO standards due to their proven track record on a global scale. It is strongly recommended for CCUS-EOR projects and holds potential for CCS initiatives. However, for widespread adoption, two key steps must be taken. Firstly, identifying the most fit-for-purpose polymer matrix, primarily epoxy grades, is essential to address oxidative impurities in anthropogenic CO_2 resources. New testing methods need to be developed to demonstrate that GRE can endure these corrosive environments for a reasonable lifespan, typically 25 years or more. Lastly, further cost reduction is imperative, achievable through large-scale manufacturing and global collaboration, to meet the requirements of developing countries. Unlike CRA, where expensive metal elements like Cr, Ni, and Mo are unavoidable, the cost of GRE liners can be significantly reduced as the majority of the expense arises from processing rather than raw materials.

In principle, CRA-clad liners, particularly those with metallurgical bonds, represent the future as they effectively combine the cost efficiency of steel tubing with the superior corrosion resistance of CRA materials. However, the current market for downhole tubulars is limited due to their high cost, which can hardly compete with solid CRAs. The cost could certainly be reduced if this technology became the focal point of the entire industry. In fact, several product lines have already been introduced for the production of CRA-lined tubulars in China through methods like welding overlay and explosive bonding. These products have initially found success in drill tubular with lower lifespan requirements and are now undergoing extensive testing in completion and CCUS injection to replace CRAs. Thanks to lower manufacturing costs in China, this technology demonstrates significant cost savings over solid CRAs, motivating the entire industry to optimize it, with more testing data becoming available. Similar to plating, CRA-lined technology is recommended for CCUS injection well applications when its cost can be justified.

Acknowledgements

The authors would like to thank the management of CNPC USA Corporation for granting them permission to publish this work.

Author details

Lei Zhao*, Jiaxiang Ren, Timothy Ryan Dunne and Peng Cheng
CNPC USA Corporation, Houston, USA

*Address all correspondence to: lei.zhao@cnpcusa.com

References

[1] Findlay P. Global CCUS outlook: Bullish and ambiguous. In: SPE-AAPG-SEG CCUS Conference, Houston. Houston, Texas: Society of Petroleum Engineers (SPE); 2023

[2] AMPP Guide 21532. Guideline for Materials Selection and Corrosion Control for CO_2 Transport and Injection. Houston, Texas: AMPP; 2023

[3] Farouq Ali S, Soliman M. Limitations and fallacies of carbon capture and storage (CCS) and impact on oil and gas production. In: SPE Annual Technical Conference and Exhibition. San Antonio. Houston, Texas: Society of Petroleum Engineers (SPE); 2023

[4] Cailly B, Thiez PL, Egermann P, Audibert A, Vidal-Gilbert S, Longaygue X. Geological storage of CO2: A state-of-the-art of injection processes and technologies. Oil & Gas Science and Technology. 2005;**60**:517-525

[5] Pimenta G, Romero C. CO_2 EOR challenges and opportunities to enable CCUS strategies. In: SPE ADIPEC, Abu Dhabi. Houston, Texas: Society of Petroleum Engineers (SPE); 2022

[6] N. SP0775. Preparation, Installation, Analysis, and Interpretation of Corrosion Coupons in Oilfield Operations. Houston, Texas: AMPP; 2018

[7] Nesic S, Postlethwaite J, Olsen S. An electrochemical model for prediction of corrosion of mild steel in aqueous carbon dioxide solutions. Corrosion; 1996;**52**:280-294

[8] Gulbrandsen E, Bilkova K. Solution chemistry effects on corrosion of carbon steels in presence of CO_2 and acetic acid. In: Corrosion 2006, San Diego, California. Houston, Texas: Association for Materials Protection and Performance (AMPP); 2006

[9] Zhao L, Ren J, Dunne T, Li J, Cheng P, Qi C. Cost effective anti-corrosion strategies for replacing corrosion-resistant alloys CRA in CCUS injection Wells. In: Offshore Technology Conference, Houston. Houston, Texas: Society of Petroleum Engineers (SPE); 2024

[10] Anderko A, Cao L, Gui F, Sridhar N, Engelhardt G. Modeling localized corrosion of corrosion-resistant alloys in oil and gas production environments: Part II. Corrosion potential. Corrosion. 2017;**73**:634-647

[11] Munger CG, Vincent LD. Corrosion Prevention by Protective Coatings. Houston, Texas: AMPP; 2014

[12] Paul S, Shepherd R, Woollin P, Park G. Selection of materials for high pressure CO_2 transport. In: Third International Forum on the Transportation of CO_2 by Pipeline, Newcastle. New York, NY: American Society of Mechanical Engineers (ASME); 2012

[13] Winter M, Mukherjee S, Datta V, O'Donoghue M. Tank linings for high temperature services. In: Corrosion 2012, Salt Lake City. Houston, Texas: Association for Materials Protection and Performance (AMPP); 2012

[14] Bahadori A. Essentials of Coating, Painting, and Lining for the Oil, Gas and Petrochemical Industries. Philadelphia: Elsevier; 2015

[15] Groves S, Wood R, Mellor B, Symonds N. Impact testing of downhole tubing internal coatings. In: Corrosion

2001, Houston. Houston, Texas: Association for Materials Protection and Performance (AMPP); 2001

[16] Gill T. Ten years of handling CO2 for SACROC unit. In: SPE Annual Meeting, New Orleans. Houston, Texas: Society of Petroleum Engineers (SPE); 1982

[17] NACE TM0185. Evaluation of Internal Plastic Coatings for Corrosion Control of Tubular Goods by Autoclave Testing. Houston, Texas: AMPP; 2006

[18] TM 0186. Holiday Detection of Internal Tubular Coatings of 250 to 760 μm (10 to 30 Mils) Dry-Film Thickness. Houston, Texas: AMPP; 2016

[19] NACE TM0183. Evaluation of Internal Plastic Coatings for Corrosion Control of Tubular Goods in an Aqueous Flowing Environment. Houston, Texas: AMPP; 2018

[20] NACE SP0291. Care, Handling, and Installation of Internally Plastic-Coated Oilfield Tubular Goods and Accessories. Houston, Texas: AMPP; 2023

[21] ASTM D2794. Standard Test Method for Resistance of Organic Coatings to the Effects of Rapid Deformation (Impact Resistance). West Conshohocken: ASTM; 2019

[22] Wright I, Ringrose P, Mathieson A, Eiken O. An overview of active large-scale CO_2 storage projects. In: SPE International Conference on CO_2 Capture, Storage, and Utilization, San Diego. Houston, Texas: Society of Petroleum Engineers (SPE); 2009

[23] Smith L, Billingham M, Lee C-H, Milanovic D. Establishing and maintaining the integrity of Wells used for sequestration of CO_2. Energy Procedia. 2011;**4**:5154-5161

[24] Maclntyre K. Design considerations for carbon dioxide injection facilities. The Journal of Canadian Petroleum Technology. 1985. pp. 90-95

[25] IEAGHG. Corrosion and Selection of Materials for Carbon Capture and Storage. United Kingdom: IEAGHG; 2010. Available from: https://ieaghg.org/publications/corrosion-and-selection-of-materials-for-carbon-capture-and-storage/

[26] Smith L. Control of corrosion in oil and gas production tubing. British Corrosion Journal. 1999;**34**(4):247-253

[27] Houghton CJ, Westermark RV. North Sea downhole corrosion: Identifying the problem; implementing the solutions. Journal of Petroleum Technology. 1983;**35**:239-246

[28] Zhang L. Corrosion in CO_2 Capture, Transportation, Geological Utilization and Storage. Berlin: Springer; 2023

[29] Zhu D, Uzcategui A, Gong L, Qiu X, Huang J, Huang J, et al. Investigation of Electroless nickel-phosphorus coating as an alternative to corrosion/fouling resistant alloys in downhole service. In: Offshore Technology Conference Brazil, Rio de Janeiro, Brazil. Houston: Society of Petroleum Engineers (SPE); 2017

[30] Sun C, Zeng H, Luo J. Unraveling the effects of CO_2 and H_2S on the corrosion behavior of electroless Ni-P coating in $CO_2/H_2S/Cl^-$ Environments at high temperature and high pressure. Corrosion Science. 2019;**148**:317-330

[31] Zhu L, Xie B, Li G, Huang X, Feng C, Lu C, et al. Damage reason of tubing with Ni-P plating in water injection wells of an oilfield. Corrosion and Protection. 2020;**41**:63

[32] Guo G, Li Q, Cheng B, Lu W, Yang L, Zhou Z. Electrode tool logging approaches for casing inner coating monitoring. In: Corrosion 2020, Vitual. Houston, Texas: Association for Materials Protection and Performance (AMPP); 2020

[33] Ma Q, Liang G, Guo Y, Sun H, Chen C. Corrosion behavior of tungsten alloy coating in the underground environment of oil field. Surface Technology. 2019;**48**:78-84

[34] Chen T. Research and application of anticorrosion technology in y-CO_2 injection test area of Daqing oilfield. Total Corrosion Control. 2021;**35**:135-138

[35] Fuzhou W, Gao C, Yu L. Anti-corrosion technique of injection well in CO_2 flooding. In: SPE International Conference on Oilfield Corrosion, Aberdeen. Houston, Texas: Society of Petroleum Engineers (SPE); 2010

[36] Davis RH, Snider B. Successful Injection, Disposal, and Production Well Applications of Thermoplastic Lined Downhole Tubing: A Compilation of Case Histories Dating Back to 1996. Humble, Texas: Western Falcon, Inc.

[37] Sharif QJ, Esmail OJ, Radhakrishnan G, Simpson JA, Bremner MR. Experience with fiberglass (GRE) lined carbon steel tubular for corrosion protection for oil production applications. In: in the SPE Annual Technical Conference and Exhibition, San Antonio. Houston, Texas: Society of Petroleum Engineers (SPE); 2012

[38] API RP 15CLT. Recommended Practice for Composite Lined Steel Tubular Goods. Washington: American Petroleum Institute (API); 2007

[39] Radhakrishnan G, Sharma ALK, Venkataraman E, Fernandes A. Enhanced life cycle costing approach to tubing material selection - New cost and revenue drivers identified from experience with non-Metallics downhole. In: Abu Dhabi International Petroleum Exhibition & Conference, Abu Dhabi, UAE. Houston, Texas: Society of Petroleum Engineers (SPE); 2018

[40] Waterton K, Savino V, Radhakrishnan G, Simpson J. Application of glass reinforced epoxy (GRE) lining barriers in downhole tubing for gas production under sweet conditions. United Kingdom: Intertek-CAPCIS

[41] Melve B, Nice PI, Axelsen SB, Waterton K. Qualification of Composite Material Lined Downhole Tubing for Gas Production. Trondheim, Norway: Statoil ASA

[42] Ross K. GRE composite-lined tubular products in corrosive service: A study in workover economics. In: SPE Permian Basin Oil and Gas Recovery, Midland, Texas. Houston, Texas: Society of Petroleum Engineers (SPE); 2001

[43] Ross K. GRE composite-lined tubular products in corrosive service: A study in workover economics. In: SPE Permian Basin Oil and Gas Recovery, Midland. Houston, Texas: Society of Petroleum Engineers (SPE); 2001

[44] Guerra D. Usage of Corrosion Resistant Alloys –CRA –for Oil and Gas Production (OCTG). Vallourec; 2016

[45] Fitzgerald A, Groves S, Gosch S, Morey S, Pattillo D, Piskurich N, et al. Alternatives to 25% chrome for water-injection tubing in deep water. In: 2008 SPE Annual Technical Conference and Exhibition, Denver. Houston, Texas: Society of Petroleum Engineers (SPE); 2008

[46] Smith L, Milanovic D, Lee CH, Billingham M. Establishing and

maintaining the integrity of Wells used for sequestration of CO_2. In: Corrosion 2012, Salt Lake City, Utah. Houston, Texas: Association for Materials Protection and Performance (AMPP); 2012

[47] Hughes D. Oil & gas - maximising recovery programme (formerly SHARP) IOR views. In: Twelfth Annual CO_2 Flooding Conference. Midland, Texas: Midland CO2 Conference; 2006. Available from: https://co2conference.net/

[48] Macintyre K. Design considerations for carbon dioxide injection facilities. Journal of Canadian Petroleum Technology. 1986;**25**:90-95

[49] Li X, Lv X, Zhongyun L, Sun X. Trial and application of modifier glass lined tube. Oil Drilling & Production Technology. 2001;**23**:76-78

[50] Nygaard R. Well Design and Well Integrity. Energy and Environmental Systems Group. Calgary, Canada: University of Calgary

[51] Underground Injection Control (UIC) Program Class VI Well Construction Guidance. Washington, D.C.: United States Environmental Protection Agency (EPA); 2012

[52] Noordin MFI, Berok SMJ, Sinanan HB, Suratman MF, Fabian O, Mohammad M, et al. Successful deployment of 15Cr OCTG CRA injection tubing in immiscible water-alternating-gas iWAG Wells - first in the world. In: International Petroleum Technology Conference, Virtual. Houston, Texas: Society of Petroleum Engineers (SPE); 2021

[53] Orlov MV, Badeghaish WO. Advanced non-metallic coatings and composite materials for O & G Industry. Journal of Physics: Conference Series. 2021;**1990**:012001

[54] Badeghaish W, Noui-Mehidi M, Salazar O. The future of nonmetallic composite materials in upstream applications. In: SPE Gas & Oil Technology Showcase and Conference, Dubai. Houston, Texas: Society of Petroleum Engineers (SPE); 2019

[55] Millet C, Lima L, Néel G, Barrera DR. Material selection for anthropogenic CO_2 injection: Mechanical and corrosion performances of steels under dense CO_2 with impurities. In: AMPP Annual Conference and Expo, Denver. Houston, Texas: Association for Materials Protection and Performance (AMPP); 2023

[56] Smith L. Engineering with CLAD Steel. North Carolina, USA: Nickel Institute; 2012

[57] Urband BE. CRA clad downhole tubing - an economical enabling technology. In: AADE 2001 National Drilling Conference, Houston. Houston, Texas: Society of Petroleum Engineers (SPE); 2001

[58] Johnston C, Nageswaran C, London T. Investigations into the fatigue strength of CRA lined pipe. In: Offshore Technology Conference, Houston, Texas, USA. Houston, Texas: Society of Petroleum Engineers (SPE); 2016

[59] Venås A, Tronskar JP. Experience with lined and clad pipelines. In: NACE Milano Italia Section - Conference & Expo 2018, Genoa. Houston, Texas: Association for Materials Protection and Performance (AMPP); 2018

[60] API Specification 5LD. CRA Clad or Lined Steel Pipe. Washington, DC: American Petroleum Institue (API); 2015

[61] Kane RD, Wilhelm SM, Yoshida T, Matsui S, Iwase T. Analysis of bimetallic pipe for sour service. SPE Production Engineering. 1991;**6**:291-296

Chapter 5

Application of Surface Engineering in Metal and Plastic Joining of Insert Injection Molding

Wen Shuai, Jialiang Qi and Jian Wang

Abstract

Insert injection molding is a widely adopted manufacturing process that integrates metal inserts into plastic components, enhancing the functionality and mechanical properties of the final product. Despite its advantages, the process often faces challenges related to the adhesion and bonding strength between metal and plastic. This chapter delves into applying surface engineering techniques to address these challenges. The adhesion between metal and plastic can be significantly improved by modifying the surface properties of metal inserts through various engineering methods such as surface treatments, coatings, and texturing. The chapter explores different surface engineering techniques and their mechanisms and presents their practical applications. Case studies will be presented to illustrate the effectiveness of these techniques in real-world scenarios, providing valuable insights for researchers and engineers in the field and promoting the advancement of hybrid material systems in manufacturing.

Keywords: surface engineering, insert injection molding, metal-plastic joining, adhesion, roughness

1. Introduction

In contemporary industrial development, lightweight design has become an increasingly prevalent demand, particularly within numerous sectors such as automotive, electronics, and household appliances [1]. This trend has catalyzed the rapid evolution and innovation of metal-plastic joining technologies. To achieve an efficient combination of materials, researchers have explored and developed various technical methods to perfectly integrate the high strength of metals with the lightweight characteristics of plastics. Among the many joining techniques, insert injection molding technology has distinguished itself due to its exceptional performance and broad applicability. It is a process where a polymer melts around the insert and is placed within the injection mold cavity, forming a single adhesive bond between the insert and the injected polymer [2]. This technology aids in fabricating robust, durable, and lightweight parts, which explains its adoption across various industries.

However, the insufficient adhesion between metals and plastics may compromise the overall mechanical performance of the product. The typically different coefficients of thermal expansion between metals and plastics can lead to internal stresses during temperature fluctuations, affecting the stability and durability of the connection [3]. Moreover, the chapter emphasizes the crucial need for carefully selecting suitable plastics and metal materials to ensure chemical and physical compatibility, as improper material selection may result in inadequate bonding or performance degradation.

Traditional methods like mechanical fastening and adhesive bonding face limitations in polymer-metal hybrid connections, constraining their application in contemporary mass production settings [4]. For instance, mechanical fastening relies on fasteners such as bolts, screws, and rivets, which not only increases the tools and processing steps in the production process but may also cause material damage and galvanic corrosion in the connection area. Additionally, using fasteners can compromise the structural integrity of the components and induce stress concentration at the connection points, thereby affecting the overall stability and durability of the structure. Conversely, adhesive bonding requires crucial surface treatment for any adhesive and additional curing time. It also faces the consequence that structural adhesives' bonding strength may decrease under high temperatures and humidity [5]. Moreover, adhesive materials may lead to thermal degradation and the emission of harmful gases. Therefore, direct connections that do not utilize additional parts or adhesives have garnered widespread attention.

Directly connecting polymers and metal components through injection molding is typically unattainable due to the inability to achieve or only the capability to attain inadequate adhesion [6]. Consequently, selecting the appropriate surface treatment method is of critical importance. Surface engineering technology is pivotal in addressing adhesion issues, significantly enhancing the connection performance between metals and plastics by altering the surfaces' physical, chemical, and morphological properties. With the advancement of new materials and technologies, insert injection molding technology has proven to be an effective solution in manufacturing automotive components, encapsulation of electronic components, and structural design of household appliances [7]. This chapter will focus on analyzing the application of surface engineering in metal and plastic injection-molded inserts, laying the groundwork for future research in related fields.

2. Fundamentals of insert injection molding

The history of metal insert plastic injection molding dates back to the early twentieth century when initial attempts were made to combine metal and plastic in manufacturing. However, it was not until the 1950s that significant advancements were achieved in this field [8]. As depicted in **Figure 1**, the formation of products in the insert injection molding process can be delineated into three critical steps. Initially, the metal insert is positioned within the mold, with the design ensuring the correct orientation and positioning of the insert within the mold. The loading methods are categorized into automatic and manual insertion; the former offers high efficiency but at a higher cost, while the latter is more cost-effective, potentially compromising precision and efficiency. In actual production, the choice depends on the complexity of the components and the production requirements. Subsequently, molten plastic is

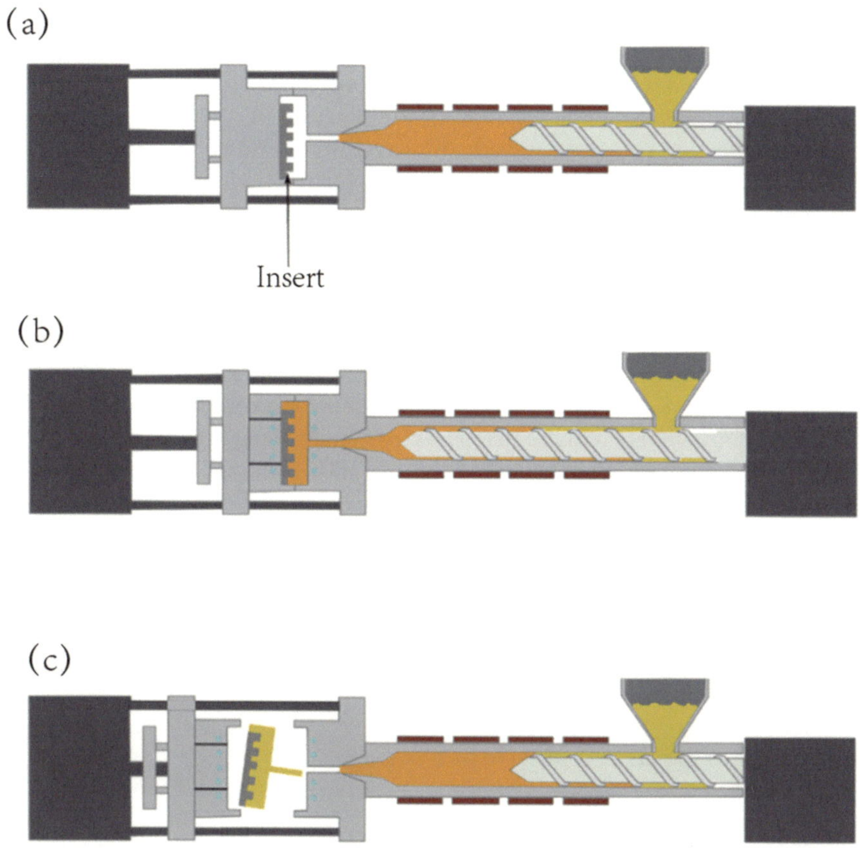

Figure 1.
Insert injection molding process: (a) Dosage and plasticzing; (b) injection; (c) cooling and ejection [5].

injected into the mold under high pressure using an injection unit, ensuring that the plastic fills the mold and uniformly adheres to the insert [9]. The final stage involves cooling and solidification, maintaining the mold at a set temperature to allow for even plastic solidification, with continued pressure to minimize shrinkage effects. After cooling, the mold is opened to eject the formed component.

Common types of metal inserts include threaded inserts, heat stake inserts, compression inserts, and expansion inserts, the choice of which is contingent upon the specific requirements and applications of the part [10]. For instance, threaded inserts, as one of the frequently utilized metal inserts in injection molding, possess external threads that can be screwed into the plastic during the molding process, thereby providing a secure and reliable connection. They are suitable for scenarios where threaded functionality needs to be incorporated into plastic components for subsequent connection with screws or bolts. Heat stake inserts may be selected when a permanent attachment of metal parts to plastic is desired. These inserts feature a flange or head that forms a sturdy bond between the metal and plastic when heated and pressed into the plastic. In cases where plastic components require metal parts that can be easily detached or replaced, compression inserts are considered. Designed with a compressible structure, these inserts are inserted into the plastic during molding and then expand to create a tight fit. Expansion inserts are utilized in applications where the shrinkage of the plastic is used to secure the insert, which is ideal for situations requiring a securing force without the need for tightening with external force. The insert is designed with a unique shape or grooves to accommodate the shrinkage of the plastic, thereby achieving fixation. In summary, each type of insert possesses its specific application scenarios and advantages; the selection of the appropriate insert type hinges on the design requirements, performance demands, and cost-effectiveness considerations of the final product.

In the modern manufacturing industry, particularly in the automotive and aerospace sectors, the technology of metal-plastic injection-molded inserts is gradually replacing traditional all-steel structural components due to its exceptional lightweight characteristics and high integration capability. This technology has demonstrated outstanding performance in the automotive manufacturing sector, being utilized in critical components such as bumper beams, dashboard crossbeams, and front-end modules. It effectively reduces material costs while being highly favored in the industry for its superior load-bearing capabilities [11]. The importance of metal-plastic injection-molded insert technology in modern manufacturing is reflected in its enhancement of product performance and reliability and the unprecedented flexibility and cost-effectiveness it brings to product design and manufacturing processes. This technology optimizes material performance by combining metals' high strength with plastics' lightweight properties. It reduces costs, providing strong support for the sustainable development of the manufacturing industry.

3. Challenges in metal and plastic joining

3.1 Adhesion problem

Given the substantial differences in physicochemical properties between plastics and metals, achieving high-quality interfacial adhesion presents a formidable challenge [12]. Polymers are typically characterized by their lower surface energy and adhesion strength than metallic materials. Consequently, to facilitate a beneficial union between metals and plastics, implementing various surface treatments on polymers to enhance their adhesive properties is deemed essential [13]. Many factors influence adhesive strength in connecting metals and plastics, a critical performance indicator. For instance, the surface roughness of metals and plastics can affect the mechanical interlocking effect of the bond [14]. Generally speaking, the rougher the surface, the higher the adhesive strength. The enhancement of surface roughness on metals facilitates the infiltration of polymer melt into micro-grooves, thereby expanding the contact area and enhancing the adhesive strength [15]. Additionally, the chemical structure of the metal and plastic surfaces, such as functional groups, significantly impacts the adhesive strength. Surface treatment techniques, such as plasma and chemical processing, can introduce or enhance these chemical structures. As shown in **Figure 2**, utilizing surface treatment methods, either mechanical or chemical, to generate micro and nano-scale textures on metallic surfaces can enhance adhesion. The molten polymer fills these textured surfaces during the injection molding process, leading to a strong bond. The materials' thermal expansion coefficients also play a role in the adhesive strength. The contrast in thermal expansion coefficients between metals and plastics can lead to internal stresses during temperature changes, affecting the adhesive strength. Selecting materials with matched thermal expansion coefficients or designs can mitigate this impact [16].

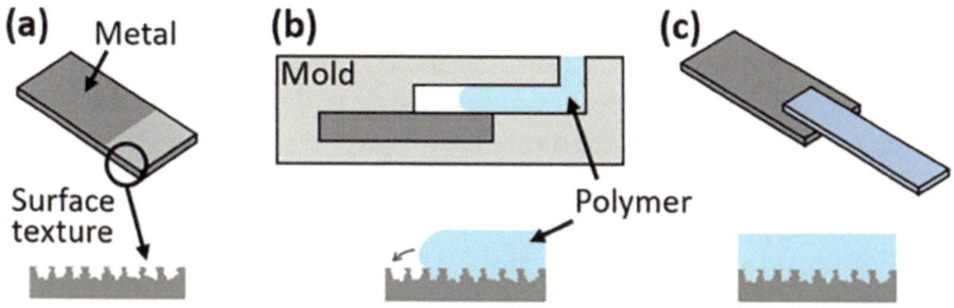

Figure 2.
Surface treatment technology: (a) Treatment of metal surface to form texture; (b) the molten polymer is injected into the treated metal surface by injection molding; (c) formation of metal–polymer hybrid joint after solidification of polymer [9].

3.2 Mechanisms for enhancing adhesion

Since the early days of the aerospace and automotive industries, adhesion mechanisms have been recognized as closely related to the surface properties of materials, involving interactions between atoms and molecules at the interface [17]. Atomic interactions between metals and polymers at the bonding interface are primarily categorized into physical adsorption and chemical bonding [18]. Various physical interactions in mechanical engineering contribute to improving interfacial adhesion between polymers and metals. Notably, van der Waals forces and hydrogen bonding are fundamental. Van der Waals forces encompass London interactions, Keesom interactions, and Debye interactions, the weakest interactions between atoms in contact. Hydrogen bonds, relatively more robust than van der Waals forces, can form at the polymer-metal interface due to interactions between hydrogen atoms on the polymer and metal surfaces, significantly contributing to the adhesion strength. However, hydrogen bonds are considerably weaker compared to covalent bonds. Collectively, these interactions enhance the adhesive strength at the interface between polymers and metals, which in turn influences the bonding performance and the overall properties of the materials.

Recently, numerous researchers have sought to enhance adhesive strength by optimizing the mechanical interlocking effect. This has been achieved by creating deeper surface structures that combine micro and nano-scale features, thereby improving the interfacial integration and overall bonding strength [19]. Research indicates that an appropriate surface structure on metal surfaces, such as grooves, pores, or protrusions, can significantly improve the adhesion between metal and polymer-based materials by creating micro-mechanical interlocks at the interface [20]. For instance, Chen and colleagues suggested a secondary heat treatment to strengthen the bond between sandblasted steel and the polymer matrix [21]. Initially, SPCC (Steel Plate Cold Rolled Commercial Quality) is subjected to sandblasting treatment, ultrasonic cleaning, and heat treatment at 500°C. Subsequently, the joint is created through injection molding. During this process, sandblasting treatment enhances the surface roughness, creating a micro-scale rough texture. The roughness of metal and plastic surfaces can influence the mechanical interlocking effect of adhesion, thereby increasing the bonding strength. Moreover, the additional heat treatment further enhances the surface energy and wettability, forming nanostructures and significantly improving the adhesive strength at the polymer-steel interface.

Furthermore, integrating functional chemical groups that can chemically bond with the plastic significantly bolsters the adhesion between metal and plastic materials. For instance, Zhao and colleagues have created metal-plastic hybrid joints by injection molding onto the aluminum surface to facilitate the adsorption of water [22]. Firstly, the aluminum plate is subjected to alkaline and acid cleaning to remove the original oxidized pollutants, followed by rinsing in deionized water. Subsequently, the aluminum plate undergoes a boehmite treatment involving immersion in deionized water at 95°C for 5 minutes, forming an amorphous AlOOH structure. Finally, a heat treatment is applied to minimize surface water absorption, which helps standardize the aluminum surface's chemical state and provides a uniform surface for subsequent bonding. These surface engineering techniques can significantly enhance the bonding strength, durability, and reliability between metal and plastic, meeting the demands of various industrial applications.

4. Surface engineering technology

4.1 Definition

Surface engineering is applying various methods and processes to improve the strength, durability, and performance of the connection between two materials by altering the chemical, physical, and morphological attributes of the surfaces of metals or plastics [23]. This typically involves surface cleaning, activation, modification, or coating to augment the adhesive force, mechanical interlocking, or chemical bonding between metals and plastics. It is an interdisciplinary field that extends beyond the surface science of materials to include various disciplines such as physics, chemistry, mechanical engineering, electronic engineering, and bioengineering. Through surface engineering, the performance of materials can be significantly improved, their service life extended, or new functionalities endowed to meet the demands of specific applications.

According to literature reports, surface engineering techniques applicable to metal and plastic injection-molded inserts can be categorized into three major classes: physical techniques like abrasive blasting, peening, and wear processes, alongside chemical treatments including cleaning, surface etching, adhesion promoters, and electrolytic oxidation, additionally, leveraging energy-based techniques such as plasma treatment, laser processing, and radiation exposure [24].

Physical techniques modify the material's surface by eliminating the existing protective coating and creating a textured surface. This process improves the adhesion between the layers by promoting mechanical interlocking and achieving a cleaner surface [25]. Lucchetta et al. subjected 6062 aluminum alloy plates to shot peening with varying sand grain diameters to elevate the surface texture of the initial aluminum sheets, followed by injection molding [26]. Kajihara et al. proposed using sandblasting to create surface textures on metal surfaces. By altering the type of abrasive, particle size, blasting pressure, and scanning speed, surface textures of different dimensions and densities can be formed. To enhance the strength of the joints, the conditions for blasting and molding have been optimized, achieving strong bonding without needing a metal preheating process [27]. As illustrated in **Figure 3**, the fiber content and roughness increase significantly enhances the adhesive strength at the polymer-metal interface.

Chemical techniques modify the chemical characteristics of a surface by eliminating contaminants, introducing reactive functional groups, and creating chemical bonds; the interfacial adhesion is fortified. Zhao and colleagues have demonstrated that in direct joints of aluminum/polyamide 6 (PA6) manufactured by direct injection

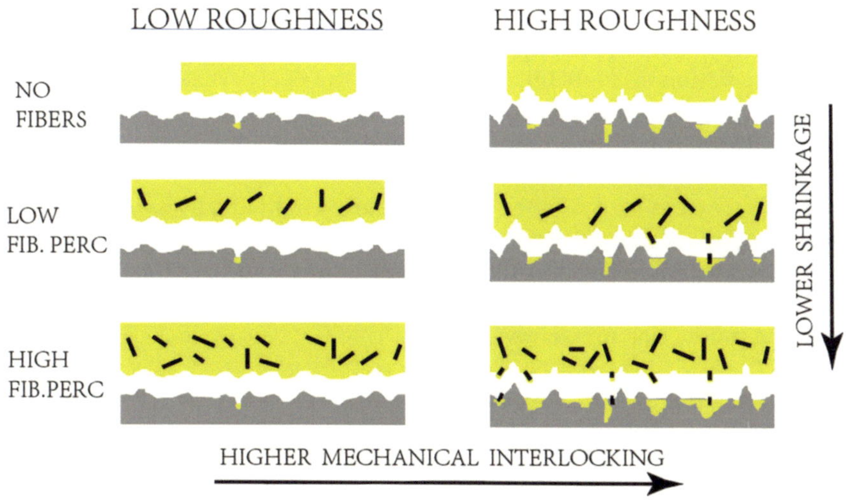

Figure 3.
Polymer-metal interlocking phenomenon [27].

molding, hydrogen bonds form between the –CONH in PA6 and the hydroxyl groups (–OH) on the aluminum surface, which strengthens the bonding affinity between metallic and polymeric materials. As shown in **Figure 4**, polar groups/molecules (such as hydroxyl –OH) are introduced to the surface of the aluminum material through hot water treatment. Then, the number of surface polar groups/molecules is altered by heat treatment [28]. Shanmugam and co-researchers strengthened the interfacial adhesion of Ti6Al4V with polyethylene fiber-reinforced thermoplastic polymer composites through the process of anodization, followed by etching and annealing post-treatments, increasing the type I interlaminar fracture toughness from 0.25 to 1.57 kJ/m^2 [29].

Energy methods, which utilize high-energy beams for surface roughening and chemical modification, offer more effective means to address weak interfacial bonding issues. Plasma and laser surface treatments are the predominant energy-based techniques to strengthen the interfacial bonding between metallic and polymeric materials. Drummer and colleagues incorporated a plasma etching pretreatment step into the injection molding process, utilizing a robotic arm with six degrees of freedom

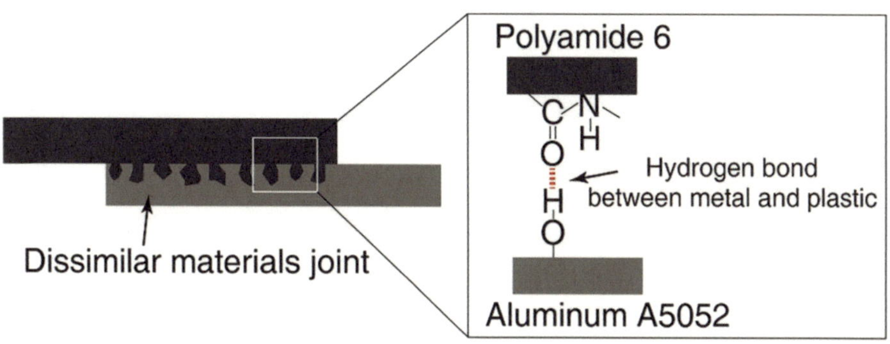

Figure 4.
Mechanism of chemical modification [28].

to maneuver the plasma nozzle [30]. Adjusting the metal preheating temperature and mold temperature can improve the adhesion strength. Rodriguez-Vidal et al. have utilized laser irradiation to create micro-patterns on the steel surface and manufactured metal-polymer interfaces using a laser emitter. The joints that incorporate micro-patterns exhibit an enhanced strength compared to those without such patterns, as depicted in **Figure 5** [31].

These surface treatment technologies can be used individually or in combination to achieve the desired surface properties. When selecting and applying these technologies, it is essential to consider the categories of metal and plastic, the expected application conditions, cost-effectiveness, environmental impact, and compatibility with existing production processes.

4.2 Case studies and practical applications

Surface engineering has been applied to insert injection molding in numerous case studies. Verma et al. [32] designed a crystal lattice structure with numerous curved pillars on stainless steel (SUS 316L) substrates, utilizing 3D Printing technology to achieve superior tensile properties. Afterward, the resin was infused into the lattice structures' pores using the injection molding method to strengthen the bond at the metal-resin interface. In the injection molding process, the molten resin penetrates the lattice's interstices and, upon solidification, forms a strong mechanical interlock, significantly enhancing the strength of the metal-plastic adhesion. As the 3D-printed lattice frameworks can be optimized according to design freedom to accommodate specific load conditions and application requirements, these structures are anticipated to exhibit enhanced durability. The lattice structure design allows for better distribution and bearing of loads, thereby reducing the risk of material fatigue and failure. However, flaws or air pockets that emerge within the 3D printing phase may undermine the strength of the material's adhesion. The interfacial integrity between the lattice and the resin is anticipated by employing Finite Element Analysis (FEA) techniques. It is correlated with experimental results, which can reasonably predict the failure strength of the bond. This predictive capability enhances the reliability of structural design.

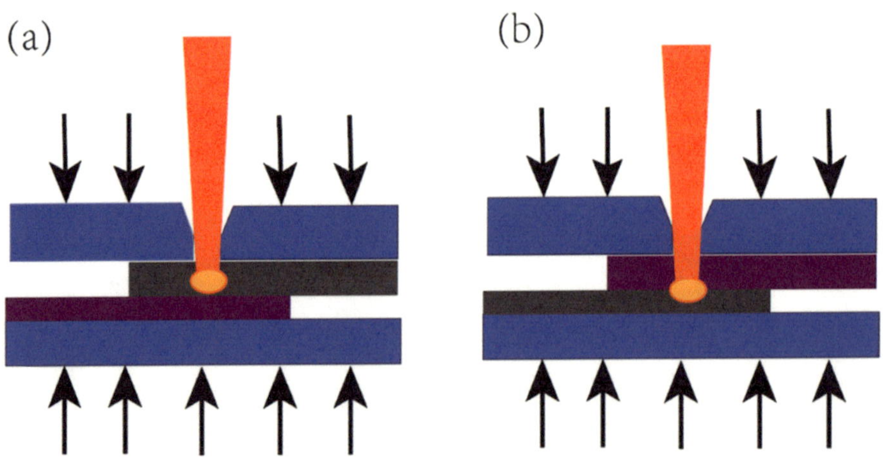

Figure 5.
Schematic diagram of laser processing method. (a) Heat conduction joining; (b) laser transmission joining [31].

Li et al. [33] placed metal substrates treated with sandblasting technology on the mold surface, and then injected the plastic melt into the metal surface features through an injection molding machine to form polymer-metal composite (PMC) parts. They utilized sandblasting technology to treat the metal surface, obtaining surface features with different height differences by altering the pressure level and grit dimension throughout the sandblasting procedure. After sandblasting, the metal exterior underwent a thorough cleaning to remove impurities. The optimization of surface features is conducive to enhancing the interfacial engagement between the molten polymer and the metallic substrate, thereby enhancing the adhesion strength. Increasing the metal surface temperature helps reduce the polymer melt's consistency and to improve its flowability. Fluidity makes the melt easier to fill the metal surface features and enhances adhesion strength.

Jiao et al. [34] utilized pure copper (C1100), reinforced polybutylene terephthalate, and polyphenylene sulfide to fabricate joint structures. Before the injection molding, the copper substrate underwent chemical etching, electrochemical deposition, and air plasma treatment to augment the number of polar groups. Scanning Electron Microscopy (SEM) imaging exposed that the polymer infiltrated the nanoscopic pores of the copper, with this interlocking effect being pivotal for enhancing adhesion strength. Molecular dynamics simulations indicated that the PBT/copper interface demonstrated more incredible interfacial interaction energy and higher maximum pull-out force than the PPS/copper interface, suggesting a more robust adhesive force between PBT and copper.

Hirahara et al. [35] bonded metal and polymer by implanting milled glass fibers into the metal surface and injecting molten polymer onto it. Two types of glass fibers with different lengths (80 and 40 micrometers) were incorporated onto the surface of an aluminum 5052 base material, followed by injection molding with polyester polymer. The aluminum substrate was first sandblasted with typical aluminum oxide powder. During the sandblasting process, a laser pulse was continuously applied to the aluminum surface to melt the substrate surface without causing macroscopic morphological changes while implanting the glass fibers. Shear strength tests revealed that samples with implanted glass fibers exhibited shear strengths up to three times higher than those treated with sandblasting alone, with average shear strengths reaching 17.6 and 17.9 MPa, compared to 5.4 MPa for the non-implanted samples. SEM images taken after shear testing the fiber-implanted metal surfaces showed a significant amount of polymer residue, indicating strong adhesion. The presence of this adhesion contributes to the enhancement of the structural durability. The fiber implantation technique reduces reliance on chemical adhesives by forming mechanical anchors on the exterior of the metal, thereby enhancing the bond's dependability. Additionally, the interfacial bonding performance can be further optimized by controlling the depth and orientation of the fiber implantation, strengthening the overall structural reliability.

Li et al. [36] employed the electrostatic spray application method to create a thermoplastic coating on the metal surface. Subsequently, polymer composites were injected directly into molds on the pre-sprayed metal surface to augment the adhesive bond with the composite. After cleaning, the metal surface was electrostatically sprayed with PA66 powder to create a thermoplastic coating approximately 0.1 mm thick, intended to strengthen the adhesion between the polymer composites and steel within the PMH material framework. This process leveraged a convex-concave interlocking mechanism, where the infusion of material as the injection molding proceeds formed micro-scale mechanical interlocking, thereby increasing the adhesion strength. Optimized interfacial treatment and bonding processes were employed to enhance the durability under cyclic loading.

The failure modes observed through experiments, such as interfacial delamination, matrix brittle fracture, fiber rupture or extraction, and plastic fracture of the metal, provided a basis for further optimization of material design and processing techniques, improving material reliability.

4.3 Experimental methods and characterization

The characterization of surface properties can generally be divided into two categories: the characterization of surface physical conditions and the characterization of surface chemical conditions.

Characterization of surface physical properties can be conducted in various ways. Regular methods include direct observation and examination of the micromorphology and composition of the material surface using atomic force microscopy (AFM), Transmission Electron Microscopy (TEM), and Scanning Electron Microscopy (SEM).

In characterizing the hydrophilicity and hydrophobicity of material surfaces, a contact angle measurement system is generally used to measure the contact angle between the material surface and the liquid. Given the direct relationship between the wetting angle and the texture of the surface, one can determine the hydrophilicity of the material [37], as shown in **Figure 6**. A wetting angle below 90° signifies high surface wettability [38], and the surface exhibits an affinity for water. Conversely, if the wetting angle exceeds 90°, it suggests poor wettability and the surface is hydrophobic. One method for measuring surface roughness involves using a surface profilometer. The stylus of the profilometer is placed on the clean and prepared surface to be measured, ensuring good contact between the surface and the stylus. The assessment device is activated, and the stylus moves along the surface, recording the variations in surface height. The measurement device calculates the surface roughness parameters based on the height variations along the stylus's path [10]. Optical profilometers mainly use white light interference technology, confocal microscopy technology, and laser triangulation technology to achieve non-contact surface measurements. White light interference technology reconstructs the three-dimensional topography of the surface by analyzing the interference

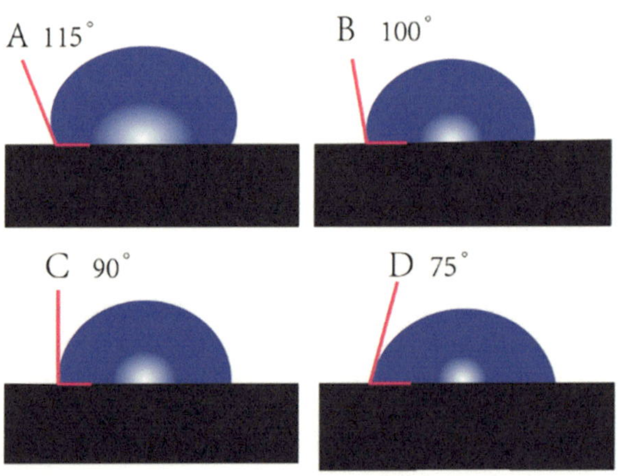

Figure 6.
Wetting angles across various surface profiles. (a) 115°; (b) 100°; (c) 90°; (d) 75° [37].

pattern of the reflected light, which is suitable for various surfaces from smooth to rough and can achieve nanometer-level measurement accuracy. Confocal microscopy technology uses conjugate confocal technology to provide excellent lateral resolution, which is suitable for large-angle topography measurement of complex structures and especially suitable for measurement of surfaces with low reflectivity. Laser triangulation measurement technology calculates the height information of each point by emitting a laser beam onto the sample surface and then receiving the reflected light with a camera, based on the principle of triangulation measurement, suitable for rapid measurement. The measurement of surface crystallinity is accomplished through X-ray Diffraction (XRD) technology. This technique assesses the content of crystalline and amorphous phases within a material. Integrating the area under each diffraction peak can influence the proportion of different crystalline phases.

One of the methods for surface chemical characterization is spectroscopic analysis, which assesses the chemical state and elemental composition on the surface. For instance, researchers such as Zhao [28] have used X-ray Photoelectron Spectroscopy (XPS) to measure the aluminum surface under various surface treatment conditions, which allows for the identification of the molar percentage of surface hydroxyl functionalities (–OH), adsorbed water (H_2O), and oxygen (O_2-). By employing AFM-IR (Atomic Force Microscopy-Infrared Spectroscopy) to measure the infrared spectrum at the boundary between aluminum and PA6, an observed enhancement in the hydrogen-bonded C=O's stretching vibration intensity relative to the non-bonded C=O's, as the analysis nears the interface. This suggests that the C=O linkage has established a hydrogen-bonding interaction with the surface hydroxyl groups of aluminum.

The prevalent method for assessing the adhesion and cohesive strength is to subject the bonded specimens to tensile shear testing. Throughout this procedure, the stress-strain curve is meticulously documented, affording an evaluation of the shear stress behavior at the juncture of metal and polymer. Utilizing the computed outcomes of adhesive normal and shear stresses, the adhesive strength is scrutinized under various adhesive lengths and manufacturing conditions [2]. After the material fracture, the microstructure of the fracture surface is observed using a microscope to analyze the fracture characteristics regarding the adhesive junction and the process of adhesive failure. Two failure theories are here: adhesive and cohesive failure [39]. Adhesive failure refers to the reduction in the bonding strength at the interface of the adhesive and the substrate, leading to the detachment of the adhesive layer from the adherend. Cohesive failure refers to the reduction in the cohesive strength inherent within the adhesive, resulting in the fracture of the adhesive itself, as shown in **Figure 7**.

Another commonly used method for testing adhesive strength is the T-peel test, a standard method for evaluating adhesive or bonding systems' performance. It is suitable for flexible materials. The test typically involves applying the adhesive to one material and then pulling the bonded sample at a particular speed perpendicular to the long edge, causing the bonding surface to separate gradually. The adhesive strength is assessed by measuring the force required during the separation process of the bonding surface. The test results are usually presented as a force-displacement curve, where the slope of the curve represents the adhesive strength.

5. Future trends and developments

Surface engineering and materials science are rapidly advancing fields that are pivotal in numerous industries. In recent years, surface engineering has witnessed

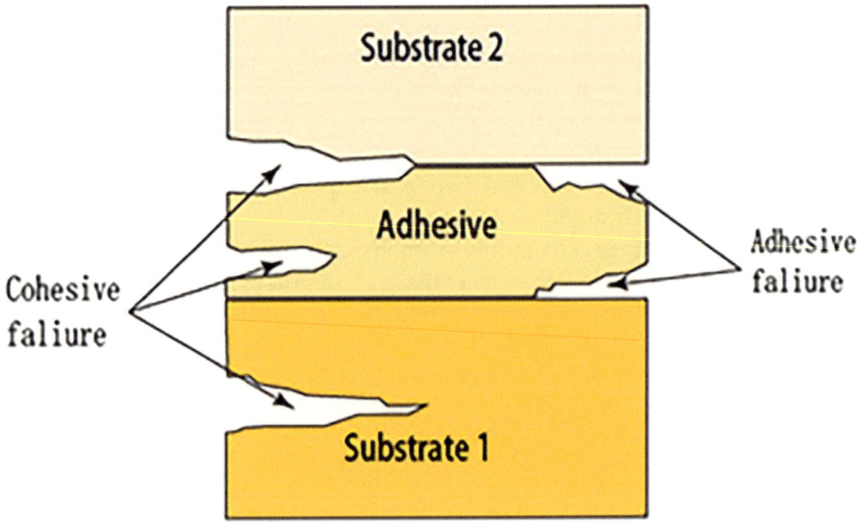

Figure 7.
Illustration of adhesive and cohesive failure [39].

the emergence of various new technologies, such as nano-coating technology that enhances surface performance using nanomaterials, laser surface treatment that alters the physical or chemical properties of material surfaces through laser technology, cold coating, and plasma treatment. For Injection Molding of Inserts (IIM), the evolution of materials science has enriched and expanded its applications. Biomaterials can be used for medical implants and in tissue engineering applications, while injection molding can fabricate high-precision medical devices using biocompatible materials. Innovative materials that respond to environmental changes, such as shape memory alloys and piezoelectric materials, may create inserts with adaptive functions. Additionally, nanomaterials and composites on a minute scale can enhance products' mechanical performance and durability.

In the future, efforts in surface engineering can focus on improving mechanical and tribological properties, developing bioactive surface treatments for medical implants, and developing corrosion-resistant coatings [40]. Additionally, intelligent surfaces can be produced using innovative materials and structures, such as shape memory alloys and piezoelectric materials, to create surfaces that respond to environmental changes. 4D printing technology integrates traditional 3D printing methods with responsive materials, enabling the creation of parts that dynamically alter their form through time.

6. Conclusion

This review discusses how surface engineering techniques significantly enhance the adhesion and connection strength between metal and plastic injection-molded inserts by altering metal and plastic surfaces' physical, chemical, and morphological properties. These methods encompass mechanical processing, chemical processing, and energy-based processing, which can enhance the physical interaction and chemical bonding between metals and plastics, thereby improving products' mechanical performance and durability. Surface engineering techniques and the connection of metals with plastics are vital to enhancing product performance and advancing the development of the manufacturing industry. These techniques can be effectively implemented by selecting appropriate surface treatment technologies, ensuring material compatibility, conducting cost-benefit analysis, and establishing strict quality control processes. At the same time, with the improvement of automation, intelligence, and environmental sustainability, future metal and plastic connection technologies will be more efficient and environmentally friendly and support customized production to meet the diverse market demands. Furthermore, integrating interdisciplinary approaches and developing innovative materials will expand the application range of metal and plastic combinations, bringing new opportunities to the manufacturing industry.

Author details

Wen Shuai[1,2], Jialiang Qi[1,2] and Jian Wang[1,2]*

1 State Key Laboratory of Organic-Inorganic Composites, Beijing University of Chemical Technology, Beijing, China

2 College of Mechanical and Electrical Engineering, Beijing University of Chemical Technology, Beijing, China

*Address all correspondence to: wjj_0107@163.com; jian.wang@buct.edu.cn

References

[1] Wu H, Qiu J, Zhang G, Sakai E, Zhao W, Tang J, et al. Interface enhancement and distribution of carbon fibers in injection welding. Polymer Composites. 2024;**45**(7):6008-6020

[2] Pinpathomrat B, Mathurosemontri S, Uawongsuwan P, Thumsorna S, Hamada H. Study on adhesive property of insert injection molded glass fiber reinforced polypropylene composites. Energy Procedia. 2016;**89**:291-298

[3] Raos G, Zappone B. Polymer adhesion: Seeking new solutions for an old problem. Macromolecules. 2021;**54**:10617-10644

[4] Chen W, Kimura F, Kajihara Y. Effect of nanostructured zinc coating on high joining strength of polymer/galvanized high-strength steel composite via injection molding. Journal of Manufacturing Processes. 2023;**85**:295-305

[5] Vasconcelos RL, Oliveira GHM, Amancio-Filho ST, Canto LB. Injection overmolding of polymer-metal hybrid structures: A review. Polymer Engineering & Science. 2023;**63**(3):691-722

[6] Rohatgi A, Sadayappan K, Clelland D, Birsan G. Joining light metals with polymer composites through metal overcasting. Journal of Materials Processing Technology. 2021;**298**:117257

[7] Blala H, Cheng P, Zhang S, et al. Advances and applications of metal-polymer hybrid for load-bearing body-in-white structural components: A comprehensive review. International Journal of Advanced Manufacturing Technology. 2024;**133**:3023-3057

[8] Zhao S, Kimura F, Kadoya S, Kajihara Y. Experimental analysis on mechanical interlocking of metal–polymer direct joining. Precision Engineering. 2020;**61**:120-125

[9] Enami K, Kimura F, Yokoyama K, Murakami T, Kajihara Y. Experimental and simulative investigation of the effects of laser-structured metal surface on metal-polymer direct joining. Precision Engineering. 2020;**62**:273-281

[10] Ramani K, Moriarty B. Thermoplastic bonding to metals via injection molding for macro-composite manufacture. Polymer Engineering and Science. 1998;**38**(5):870-876

[11] Zhang Y, Sun L, Li L, Huang B, Wang T, Wang Y. Direct injection molding and mechanical properties of high strength steel/composite hybrids. Composite Structures. 2019;**210**:70-81

[12] Zou X, Chen K, Yao H, Chen C, Lu X, Ding P, et al. Chemical reaction and bonding mechanism at the polymer–metal Interface. ACS Applied Materials & Interfaces. 2022;**14**:27383-27396

[13] Hamdi M, Saleh MN, Poulis JA. Improving the adhesion strength of polymers: Effect of surface treatments. Journal of Adhesion Science and Technology. 2020;**34**(17):1853-1870

[14] Dayss E, Leps G, Meinhardt J. Surface modification for improved adhesion of a polymer–metal compound. Surface and Coatings Technology. 1999;**116-119**:986-990

[15] Liu FC, Dong P, Pei X. A high-speed metal-to-polymer direct joining technique and underlying bonding mechanisms. Journal of Materials Processing Technology. 2020;**280**:116610

[16] Ye H, Liu XY, Hong H. Fabrication of metal matrix composites by metal injection molding—A review. Journal of Materials Processing Technology. 2008;**200**(1-3):12-24

[17] Baldan A. Adhesion phenomena in bonded joints. International Journal of Adhesion and Adhesives. 2012;**38**:95-116

[18] Lambiase F, Scipioni SI, Lee C-J, Ko D-C, Liu F. A state-of-the-art review on advanced joining processes for metal-composite and metal-polymer hybrid structures. Materials. 2021;**14**(8):1890

[19] Wang H, Jin K, Wang C, Guo X, Chend Z, Tao J. Effect of fiber surface functionalization on shear behavior at carbon fiber/epoxy interface through molecular dynamics analysis. Composites Part A: Applied Science and Manufacturing. 2019;**126**:105611

[20] Wang H, Tong X, Ji W, Chen Y, Hua L, Wu M. Improvement of interface anchoring by ultrasonic vibration for adhesively bonded CFRP/Al joints. Journal of Reinforced Plastics and Composites. 2022;**41**(1-2):20-33

[21] Chen W, Kimura F, Kajihara Y. Heat treatment induced strength improvement of the injection-molded joint between polymer and blasted steel. Materials Letters. 2023;**333**:133651

[22] Zhao S, Chen C, Liang H, Wang S, Kimura F, Kajihara Y. Investigation of the influence of aluminum surficial water on aluminum-plastic hybrids through a combined experimental and simulation approach. Applied Surface Science. 2024;**656**:159694

[23] Gopalakrishnan T, Venkatesan R, Chandrasekaran M, Deepa S, Kim S-C. Advances in implant for surface modification to enhance the interfacial bonding of shape memory alloy wires in composite resins. Progress in Organic Coatings. 2024;**188**:108242

[24] Min J, Hu J, Sun C, Liao P, Kim H. Fabrication processes of metal-fiber reinforced polymer hybrid components: A review. Advanced Composites and Hybrid Materials. 2022;**5**:651-678

[25] Kim S-G, Suzuki A, Takata N, Kobashi M. Joining of metals and polymers using powder metallurgy with laser irradiation. Journal of Materials Processing Technology. 2019;**270**:1-7

[26] Lucchetta G, Marinello F, Bariani PF. Aluminum sheet surface roughness correlation with adhesion in polymer metal hybrid overmolding. CIRP Annals - Manufacturing Technology. 2011;**60**:559-562

[27] Kajihara Y, Tamura Y, Kimura F, Suzuki G, Nakura N, Yamaguchi E. Joining strength dependence on molding conditions and surface textures in blast-assisted metal-polymer direct joining. CIRP Annals - Manufacturing Technology. 2018;**67**:591-594

[28] Zhao S, Kimura F, Wang S, Kajihara Y, Kajihara Y. Chemical interaction at the interface of metal–plastic direct joints fabricated via injection molded direct joining. Applied Surface Science. 2021;**540**:148339

[29] Zhou Z, Shan J-G, Tan X-H, Zhang J. Effect of anodizing pretreatment on laser joining CFRP to aluminum alloy A6061. International Journal of Adhesion and Adhesives. 2016;**70**:142-151

[30] Drummer D, Schmachtenberg E, Hülder G, Meister S. MK2—A novel assembly injection molding process for the combination of functional metal surfaces with polymer structures. Journal of Materials Processing Technology. 2010;**210**(3):1852-1857

[31] Rodríguez-Vidal E, Sanz C, Lambarri J, Quintana I. Experimental investigation into metal micro-patterning by laser on polymer-metal hybrid joining. Optics & Laser Technology. 2018;**104**:73-82

[32] Verma S, Yang CK, Lin CH, Jeng JY. Additive manufacturing of lattice structures for high strength mechanical interlocking of metal and resin during injection molding. Additive Manufacturing. 2022;**49**:102463

[33] Li X, Gong N, Yang C, Zeng S, Fu S, Zhang K. Aluminum polypropylene composites produced through injection molding. Journal of Materials Processing Technology. 2018;**255**:635-643

[34] Jiao Y, Ma W. Effect of the polymer on the joint strength of polymer/copper hybrids produced by nano-injection molding: Comparison of polybutylene terephthalate and polyphenylene sulfide via experimental and computational methods. Materials Today Communications. 2022;**33**:104291

[35] Hirahara T, Hidai H. Fiber implantation for interfacial joining of polymer to metal. ACS Applied Polymer Materials. 2020;**2**:3049-3053

[36] Li L, Sun L, Dai Z, Xiong Z, Huang B, Zhang Y. Experimental investigation on mechanical properties and failure mechanisms of polymer composite-metal hybrid materials processed by direct injection-molding adhesion method. Journal of Materials Processing Technology. 2019;**263**:385-395

[37] Wenzel RN. Resistance of solid surfaces to wetting by water. Industrial & Engineering Chemistry Research. 1936;**28**:988-994

[38] Zhang M, Chu L, Chen J, Qi F, Li X, Chen X, et al. Asymmetric wettability fibrous membranes: Preparation and biologic applications. Composites Part B: Engineering. 2024;**269**:111095

[39] Shrivastava A. Introduction to plastics engineering. In: Shrivastava A, editor. Plastics Design Library [M]. Norwich, NY: William Andrew Publishing; 2018. pp. 179-205. ISBN: 9780323395007

[40] Ramezani M, Mohd Ripin Z, Pasang T, Jiang C-P. Surface engineering of metals: Techniques, characterizations and applications. Metals. 2023;**13**(7):1299